PHILOSOPHY OF SCIENCE A–Z

Volumes available in the Philosophy A–Z Series

Forthcoming volumes

Philosophy of Science A–Z

Stathis Psillos

Edinburgh University Press

To my students

Edinburgh University Press Ltd
22 George Square, Edinburgh

Typeset in 10.5/13 Sabon
by TechBooks India, and printed and
bound in Great Britain by
Antony Rowe Ltd, Chippenham, Wilts

A CIP record for this book is
available from the British Library

ISBN 978 0 7486 2214 6 (hardback)
ISBN 978 0 7486 2033 3 (paperback)

Published with the support of the Edinburgh University Scholarly
Publishing Initiatives Fund.

Contents

Series Editor's Preface

Science is often seen as consisting of facts and theories, but precisely how the facts relate to the theories, and what is a fact and what is a theory have long been the subject matter of philosophy. Throughout its history scientists have raised theoretical questions that fall broadly within the purview of the philosopher, and indeed from quite early on it was not always easy to distinguish between philosophers and scientists. There has been a huge expansion of science in modern times, and the rapid development of new theories and methodologies has led to an equally rapid expansion of theoretical and especially philosophical techniques for making sense of what is taking place. One notable feature of this is the increasingly technical and specialized nature of philosophy of science in recent years. As one might expect, philosophers have been obliged to replicate to a degree the complexity of science in order to describe it from a conceptual point of view. It is the aim of Stathis Psillos in this book to explain the key terms of the vocabulary of contemporary philosophy of science. Readers should be able to use the book as with others in the series, to help them orient themselves through the subject, and every effort has been made to represent clearly and concisely its main features.

Oliver Leaman

Introduction and Acknowledgements

Philosophy of science emerged as a distinctive part of philosophy in the twentieth century. Its birthplace was continental Europe, where the neat Kantian scheme of synthetic a priori principles that were supposed to be necessary for the very possibility of experience (and of science, in general) clashed with the revolutionary changes within the sciences and mathematics at the turn of the twentieth century. The systematic study of the metaphysical and epistemological foundations of science acquired great urgency and found its formative moment in the philosophical work of a group of radical and innovative thinkers – the logical positivists – that gathered around Moritz Schlick in Vienna in the 1920s.

The central target of philosophy of science is to understand science as cognitive activity. Some of the central questions that have arisen and thoroughly been discussed are the following. What is the aim and method of science? What makes science a rational activity? What rules, if any, govern theory-change in science? How does evidence relate to theory? How do scientific theories relate to the world? How are concepts formed and how are they related to observation? What is the structure and content of major scientific concepts, such as causation, explanation, laws of nature, confirmation, theory, experiment, model, reduction and so on? These kinds of questions were originally addressed within a formal logico-mathematical framework. Philosophy of science was taken to be a largely a priori conceptual enterprise aiming to reconstruct the language of science. The naturalist turn of the 1960s challenged the privileged and foundational status of philosophy – philosophy of science was taken to be continuous with the sciences in its method and its scope. The questions above did not change. But the answers that were considered to be legitimate did – the findings of the empirical sciences, as well as the history and practice of science, were allowed to have a bearing on, perhaps even to determine, the answers to standard philosophical questions about science. In the 1980s, philosophers of science started to look more systematically into the micro-structure of individual sciences. The philosophies of the individual sciences have recently acquired a kind of unprecedented maturity and independence.

This dictionary is an attempt to offer some guidance to all those who want to acquaint themselves with some major ideas in the philosophy of science. Here you will get: concepts, debates, arguments, positions, movements and schools of thought, glimpses on the views and contribution of important thinkers. The space for each entry is limited; but cross-referencing (indicated in boldface) is extensive. The readers are heartily encouraged to meander through the long paths that connect with others the entries they are interested in – they will get, I hope, a fuller explanation and exploration of exciting and important topics. They will also get, I hope, a sense of the depth of the issues dealt with. The entries try to put the topic under discussion in perspective. What is it about? Why is it important? What kinds of debate are about it? What has been its historical development? How is it connected with other topics? What are the open issues? But the dictionary as a whole is not meant to replace the serious study of books and papers. Nothing can substitute for the careful, patient and focused study of a good book or paper. If this dictionary inspires a few readers to work their way through some books, I will consider it a success.

In writing the dictionary, I faced the difficulty of having to decide which contemporary figures I should include with separate entries. Well, my decision – after some advice – was partly conventional. Only very eminent figures in the profession who were born before the end of the Second World War were allotted entries. I apologise in advance if I have offended anyone by not having an entry on her/him. But life is all about decisions.

Many thanks are due to Oliver Leaman for his invitation to write this book; to the staff at Edinburgh University Press (especially to Carol Macdonald) for their patience and help; to Peter Andrews who copy-edited this book with care; and to my student Milena Ivanova for her help in the final stages of preparing the manuscript. Many thanks to my wife, Athena, and my daughter, Demetra, for making my life a pleasure and to my colleagues and students – who have made my intellectual life a pleasure.

Athens
May 2006

Note on Notation

Using some technical notation has become almost inevitable in philosophy. I have attempted to explain all symbols that appear in the entries when they occur, but here is a list of the most frequent of them.

&	logical conjunction
or	logical disjunction
if . . . then . . .	material conditional
if and only if (occasionally abbreviated as *iff* and as ←→)	material bi-conditional
− (occasionally *not-*)	logical negation
□→	counterfactual conditional
Aa (capital letter followed by small letter)	predicate *A* applies to individual *a*
prob(X)	the probability of *X*
prob(X/Y)	the probability of *X* given *Y*
>	greater than
∃	existential quantifier (there is . . .)

Philosophy of Science A–Z

A priori/a posteriori: There seem to be two ways in which the **truth** of a statement can be known or justified: independently of experience or on the basis of experience. Statements whose truth is knowable independently of (or prior to) experience are a priori, whereas statements whose truth is knowable on the basis of experience are a posteriori. On a stronger reading of the distinction, at stake is the modal status of a statement, namely, whether it is necessarily true or contingently so. **Kant** identified a priority with necessity and a posteriority with contingency. He also codified the **analytic/synthetic distinction**. He argued that there are truths that are synthetic a priori. These are the truths of arithmetic, geometry and the general principles of science, for example, the causal maxim: that each event has a cause. These are necessary truths (since they are a priori) and are required for the very possibility of experience. For Kant, a priori knowledge has the following characteristics. It is knowledge

1. universal, necessary and certain;
2. whose content is formal: it establishes conceptual connections (if analytic); it captures the form of pure intuition (if synthetic);

3. constitutive of the form of experience;
4. disconnected from the content of experience; hence, unrevisable.

Frege claimed that a statement is a priori if its proof depends only on general laws which need no, and admit of no, proof. So Frege agreed with Kant that a statement can be a priori without being analytic (e.g., geometrical truths), but, contrary to Kant, he thought that arithmetical truths, though a priori, are analytic. By denying the distinction between analytic and synthetic truths, **Quine** also denied that there can be a priori knowledge of any sort. The view that there can be no a priori knowledge has been associated with **naturalism**. Some empiricists think that though all substantive knowledge of the world stems from experience (and hence it is a posteriori), there can be a priori knowledge of analytic truths (e.g., the truths of logic and mathematics). Traditionally, the possibility of a priori knowledge of substantive truths about the world has been associated with **rationalism**.

See **Logical positivism; Reichenbach**
Further reading: Reichenbach (1921)

Abduction: Mode of reasoning which produces hypotheses such that, if true, they would explain certain phenomena. **Peirce** described it as the reasoning process which proceeds as follows: the surprising fact C is observed; but, if A were true, C would be a matter of course; hence, there is reason to suspect that A is true. Though initially Peirce thought that abduction directly *justifies* the acceptance of a hypothesis as true, later he took it to be a method for *discovering* new hypotheses. He took abduction to be the process of generation and ranking of hypotheses in terms of **plausibility**, which is followed by the derivation of predictions from them by means of deduction, and whose testing is done by means of **induction**. Recently, abduction has been taken as a code name for **inference to the best explanation**.

Further reading: Harman (1986); Lipton (2004)

Abstract entities: Entities that do not exist in **space** and **time** and are causally inert. Examples of abstract entities are **numbers**, sets, **universals** and propositions. They are contrasted to concrete entities, that is, spatio-temporal entities. They are also often contrasted to **particulars**, that is, to entities that are not **universals**. But these two contrasted classes need not coincide. Those who think that numbers are abstract objects need not take the view that numbers are universals: the typical view of mathematical **Platonism** is that numbers are abstract particulars. Those who think that properties are universals need not think that they are abstract entities. They may think, following essentially **Aristotle**, that universals exist only in particulars in space and time. Or, they may think, following Plato, that universals are essentially abstract entities, since they can exist without any spatio-temporal instances. There is substantial philosophical disagreement about whether there can be abstract entities. **Nominalism** denies their existence, while realism (about abstract entities) affirms it. The prime argument for positing abstract entities is that they are necessary for solving a number of philosophical problems, for instance, the problem of predication or the problem of reference of singular arithmetical terms or the problem of specifying the semantic content of statements. Deniers of their existence argue that positing abstract entities creates ontological problems (In what sense do they exist, if they make no causal difference?) and epistemological problems (How can they be known, if they make no causal difference?)

See **Concepts; Fictionalism, mathematical; Frege; Mill; Models; Reality**

Further reading: Hale (1987)

Abstraction: The removal, in thought, of some characteristics or features or properties of an object or a system that are not relevant to the aspects of its behaviour under study. In current philosophy of science, abstraction is distinguished from idealisation in that the latter involves approximation and simplification. Abstraction is an important element in the construction of **models**. Abstraction is also the process by which general concepts are formed out of individual instances, for example, the general concept TRIANGLE out of particular triangles or the general concept HUMAN BEING out of particular human beings. Certain features of particular objects (e.g, the weight or the sex of particular human beings) are abstracted away and are not part of the general concept. For **Aristotle,** abstraction is the process by which there is transition from the **particular** to the universal. In his radical critique of **universals, Berkeley** argued that the very process of abstraction cannot be made sense of.

Further reading: McMullin (1985)

Abstraction principles: Introduced by **Frege** in an attempt to explain our capacity to refer to **abstract entities**. He suggested that the concept of direction can be introduced as follows: (D) The direction of the line a is the same as the direction of the line b if and only if line a is parallel to line b. Lines are given in intuition and yet directions (introduced as above) are abstract entities not given in intuition. Accordingly, the concept DIRECTION is introduced by a process of intellectual activity that takes its start from intuition. (D) supplies identity-conditions for the abstract entity *direction of line*, thereby enabling us to identify an abstract object as the same again under a different description. Frege's fundamental thought was that the concept of number (and numbers as abstract entities) can be introduced by a similar abstraction principle,

namely: (N =) The number which belongs to the concept
F is the same as the number which belongs to the concept
G if and only if concept *F* can be in one–one correspondence with concept *G*. The notion of one–one correspondence is a logical relation and does not presuppose the
concept of number. Hence, the right-hand side of (N =)
does not assert something that is based on intuition or
on empirical fact. Still, (N =) states necessary and sufficient conditions for two numbers being the same; hence,
we are offered identity-conditions for the abstract entity
number.

Further reading: Fine (2002)

Acceptance: Attitude towards scientific theories introduced
by **van Fraassen**. It involves belief only in the **empirical
adequacy** of accepted theories, but stretches beyond **belief**
in expressing commitment to accepted scientific theories.
It is also the stance towards theories recommended by
Popperians. A theory is accepted if it is unrefuted and has
withstood severe testing.

Further reading: van Fraassen (1980)

Accidentally true generalisations: Generalisations that are
true, but do not express **laws of nature**. For instance,
though it is true that 'All gold cubes are smaller than one
cubic mile', and though this statement is law-like, it does
not express a law of nature. A typical way to tell whether
a generalisation is accidentally true is to examine whether
it supports **counterfactual conditionals**.

Further reading: Psillos (2002)

Achinstein, Peter (born 1935): American philosopher of science who has worked on **models, explanation, confirmation, scientific realism** and other areas. He is the author of
Particles and Waves: Historical Essays in the Philosophy

of Science (1991) and *The Book of Evidence* (2001). In his early work he defended a pragmatic approach to **explanation**. He has also argued that the type of reasoning that leads to and justifies beliefs about **unobservable entities** is based on a mixture of explanatory considerations and some 'independent warrant' for the truth of the explanatory hypothesis, which is based on inductive (causal-analogical) considerations. In recent work, he has defended a non-Bayesian theory of confirmation, based on objective epistemic probabilities, that is, probabilities that reflect the degrees of reasonableness of **belief**.

Further reading: Achinstein (2001)

Ad hocness/Ad hoc hypotheses: A hypothesis *H* (or a modification of a hypothesis) is said to be ad hoc with respect to some phenomenon *e* if either of the following two conditions is satisfied:

1. A body of background knowledge *B* entails (a description of) *e*; information about *e* is used in the construction of the theory *H* and *H* entails *e*.
2. A body of background knowledge *B* entails (a description of) *e*; *H* does not entail *e*; *H* is modified into a hypothesis *H'* such that *H'* entails *e*, and the *only* reason for this modification is the accommodation of *e* within the hypothesis.

   Alternatively, a hypothesis is ad hoc with respect to some phenomenon *e* if it is not independently testable, that is, if it does not entail any further predictions. A clear-cut case where the hypothesis is *not* ad hoc with respect to some phenomenon is when the hypothesis issues a **novel prediction**.

   See **Prediction vs accommodation**
   Further reading: Lakatos (1970); Maher (1993)

Ampliative inference: **Inference** in which the content of the conclusion exceeds (and hence amplifies) the content of the premises. A typical case of it is: 'All *observed* individuals who have the property A also have the property B; therefore (probably), All individuals who have the property A have the property B'. This is the rule of **enumerative induction**, where the conclusion of the inference is a generalisation over the individuals referred to in its premises. **Peirce** contrasted ampliative inference to explicative inference. The conclusion of an explicative inference is included in its premises, and hence contains no information that is not already, albeit implicitly, in them: the reasoning process itself merely unpacks the premises and shows what follows logically from them. Deductive inference is explicative. In contrast to it, the rules of ampliative inference do not guarantee that whenever the premises of an **argument** are true the conclusion will also be true. But this is as it should be: the conclusion of an ampliative argument is adopted on the basis that the premises offer *some* reason to accept it as probable.

See **Deductive arguments; Defeasibility; Induction, the problem of**

Further reading: Harman (1986); Salmon (1967)

Analogical reasoning: Form of **induction** based on the presence of analogies between things. If A and B are analogous (similar) in respects $R_1 \ldots R_n$ it is inductively concluded that they will be analogous in other respects; hence if A has feature R_{n+1}, it is concluded that B too will probably have feature R_{n+1}. The reliability of this kind of reasoning depends on the number of instances examined, the number and strength of positive analogies and the absence of negative analogies (dissimilarities). More generally, analogical reasoning will be reliable only if the

noted similarities are characteristic of the presence of a
natural kind.

See **Analogy**

Further reading: Holyoak and Thagard (1995)

Analogy: A relation between two systems or objects (or theories) in virtue of which one can be a **model** for the other. A
formal analogy operates on the mathematical structures
(or equations) that represent the behaviour of two systems X and Y. Sameness in the material properties of the
two systems need not be assumed, provided that the systems share mathematical **structure.** A *material* analogy
relates to sameness or similarity of **properties.** Material
analogies between two physical systems X and Y suggest
that one of the systems, say X, can be described, in certain ways and to a certain extent, from the point of view
of Y. **Hesse** classified material analogies in a tri-partite
way: (1) positive analogies, that is, properties that both
X and Y share in common; (2) negative analogies, that
is, properties with respect to which X is unlike Y; and (3)
neutral analogies, that is, properties about which we do
not yet know whether they constitute positive or negative
analogies, but which may turn out to be either of them.
The neutral analogies suggest that Y can play a *heuristic
role* in unveiling further properties of X.

Further reading: Hesse (1966)

Analytic/synthetic distinction: All true statements are supposed to be divided into two sorts: analytic and synthetic.
Analytic are those statements that are true in virtue of the
meaning of their constituent expressions, whereas synthetic are those statements that are true in virtue of extra-
linguistic facts. Though the distinction was present before
Kant, he was the first to codify it. Kant offered two criteria
of analyticity. According to the first, a subject-predicate

statement is analytic if the (meaning of the) predicate is contained in the (meaning of the) subject. According to the second (broader) criterion, a statement is analytic if it cannot be denied without contradiction. The two criteria coincided within the framework of Aristotelian logic. So the statement 'Man is a rational animal' comes out as analytic because (1) the predicate (RATIONAL ANIMAL) is part of the subject (MAN); and hence, (2) this statement cannot be denied without contradiction. Kant took logical and conceptual **truths** to be analytic and arithmetical and geometrical statements to be synthetic (partly because they fail the first criterion of analyticity). He also codified the distinction between a priori true and a posteriori true statements and claimed that there are statements (such as those of arithmetic and geometry) that are both synthetic and a priori. **Frege** took it that analytic statements are those that are proved on the basis of logical laws and definitions. He took logic to consist of analytic truths and, since he thought that mathematical truths are reduced to logical truths, he took mathematics to consist of analytic truths. Frege agreed with Kant that geometrical truths are synthetic a priori. For Frege, a statement is synthetic if its proof requires non-logical truths (for instance, the axioms of geometry). The logical positivists rejected the existence of synthetic a priori truths and took it that all and only analytic truths are knowable a priori. They thought that analytic truths are true by **definition** or **convention**: they constitute truths about language and its use. Hence they denied the essentialist doctrine that underlied the Kantian first criterion of analyticity. They took it that analytic truths are factually empty since they have no empirical content. They tied analyticity with necessity by means of their linguistic doctrine of necessity: analytic truths (and only them) are necessary. **Quine** challenged the very possibility of the distinction between analytic and synthetic statements. He noted that the explication of analyticity requires a notion of cognitive synonymy, and argued that there is no independent criterion of cognitive synonymy. He also stressed that there are no statements immune to revision; hence if 'analytic' is taken to mean 'unrevisable', there are no analytic statements. However, **Carnap** and other logical positivists had a relativised conception of analyticity. They took the analytic-synthetic distinction to be *internal* to a language and claimed that analyticity is not invariant under language-change: in radical theory-change, the analytic-synthetic distinction has to be redrawn within the successor theory. So 'being held true, come what may' is not the right **explication** of analyticity. For Carnap, analytic statements are such that: (1) it is rational to accept them within a linguistic framework; (2) rational to reject them, when the framework changes; and (3) there is some extra characteristic which all and only analytic statements share, in distinction to synthetic ones. Even if Quine's criticisms are impotent vis-à-vis (1) and (2), they are quite powerful against (3). The dual role of **correspondence rules** (they specify the meaning of theoretical terms and contribute to the factual content of the theory) made the drawing of the analytic-synthetic distinction impossible, even *within* a theory. To find a cogent explication of (3), Carnap had to reinvent the **Ramsey-sentences**.

See **A priori/a posteriori**

Further reading: Boghossian (1996); Carnap (1950a); Quine (1951)

Anti-realism see **Realism and anti-realism**

Approximate truth: A false theory (or belief) can still be approximately true, if it is close to the **truth**. For instance, the statement 'John is 1.70 metres tall' is false if John is actually 1.73 metres tall, but still approximately true. This notion has been central in the scientific realist tool-box, since it allows realists to argue that though past theories have been false they can nonetheless be deemed approximately true from the vantage point of their successors. Hence, it allows them to avoid much of the force of the **pessimistic induction**. This notion has resisted formalisation and this has made a lot of philosophers feel that it is unwarranted. Yet, it can be said that it satisfies the following platitude: for any statement 'p', 'p' is approximately true iff approximately p. This platitude shifts the burden of understanding 'approximate truth' to understanding approximation. Kindred notions are **truthlikeness** and **verisimilitude**.

Further reading: Psillos (1999)

Argument: A linguistic construction consisting of a set of premises and a conclusion and a(n) (often implicit) claim that the conclusion is suitably connected to the premises (i.e., it logically follows from them, or is made plausible, probable or justified by them). Arguments can be divided into deductive (or demonstrative) and non-deductive (non-demonstrative or ampliative).

See **Ampliative inference; Deductive arguments; Inference**

Aristotle (384–322 BCE): Greek philosopher, one of the most famous thinkers of all time. He was the founder of syllogistic logic and made profound contributions to methodology, metaphysics and ethics. His physical theory became the dominant doctrine until the Scientific Revolution. His epistemology is based on a sharp distinction between understanding the fact and understanding the *reason* why. The latter type of understanding, which characterises scientific **explanation** and scientific **knowledge**, is tied to finding the causes of the phenomena. Though both types of understanding proceed via deductive syllogism, only the latter is characteristic of science because only the latter is tied to the knowledge of causes. Aristotle observed that, besides being demonstrative, explanatory **arguments** should also be *asymmetric*: the asymmetric relation between causes and effects should be reflected in the relation between the premises and the conclusion of the explanatory arguments. For Aristotle, scientific knowledge forms a tight deductive-axiomatic system whose axioms are *first principles*. Being an empiricist, he thought that knowledge of causes has experience as its source. But experience on its own cannot lead, through **induction**, to the first principles: these are universal and necessary and state the ultimate causes. On pain of either circularity or infinite regress, the first principles themselves cannot be demonstrated either. Something besides experience and demonstration is necessary for the knowledge of first principles. This is a process of **abstraction** based on intuition, a process that reveals the essences of things, that is, the properties by virtue of which the thing is *what it is*. Aristotle distinguished between four types of causes. The material cause is 'the constituent from which something comes to be'; the formal cause is 'the formula of its essence'; the efficient cause is 'the source of the first principle of change or rest'; and the final cause is 'that for the sake of which' something happens. For instance, the material cause of a statue is its material (e.g., bronze); its formal cause is its form or shape; its efficient cause is its maker; and its final cause is the purpose for which the statue was made. These different types of a cause correspond to different answers to why-questions.

See **Bacon; Essentialism; Empiricism; Ockham, William of; Particular; Universals**

Further reading: Aristotle (1993)

Atomism: Any kind of view that posits discrete and indivisible elements (the atoms) out of which everything else is composed. Physical atomism goes back to Leucippus and Democritus (c. 460–c. 370 BCE) and claims that the ultimate elements of reality are atoms and the void.

Further reading: Pyle (1995)

Atomism, semantic: The view that the meaning of a term (or a concept) is fixed in isolation of any other term (or concept); that is, it is not determined by its place within a theoretical system, by its logical or conceptual or inferential connections with other terms. Though it gave way to semantic holism in the 1960s, **Carnap** held onto it and developed an atomistic theory of cognitive significance for theoretical terms. His idea was a theoretical **term** is meaningful *not* just in case it is part of a theory, but rather when it makes some positive contribution to the experiential output of the theory. By this move, Carnap thought he secured some distinction between significant theoretical **concepts** and meaningless metaphysical assertions that can nonetheless be tacked on to a theory (the latter making no empirical difference). Others take it that semantic atomism is grounded in the existence of nomological connections between concepts and the entities represented by them.

See **Holism, confirmational; Holism, semantic; Tacking paradox, the**

Further reading: Fodor and Lepore (1992)

Axiology: A general theory about the constraints that govern rational choice of aims and goals, for example, predictive success, **empirical adequacy, truth.** It is taken to be a supplement to normative **naturalism** in that it offers means to choose among aims that scientific methodology should strive to achieve.

Further reading: Laudan (1996)

Bacon, Francis (1561–1626): English lawyer, statesman and philosopher. In *Novum Organum* (*New Organon*, 1620), Bacon placed method at centre-stage and argued that **knowledge** begins with experience but is guided by a new method: the method of **eliminative induction**. His new method differed from **Aristotle**'s on two counts: on the nature of first principles and on the process of attaining them. According to Bacon, the Aristotelian method starts with the senses and particular objects but then flies to the first principles and derives from them further consequences. This is what Bacon called anticipation of nature. He contrasted this method with his own, which aims at an interpretation of nature: a gradual and careful ascent from the senses and particulars objects to the most general principles. He rejected **enumerative induction** as childish (since it takes account only of positive instances). His alternative proceeds in three stages. Stage 1 is experimental and natural history: a complete, or as complete as possible, recording of all instances of natural things and their effects. Here observation rules. Then at stage 2, tables of presences, absences and degrees of variation are constructed. Stage 3 is **induction**. Whatever is present when the nature under investigation is present or absent when this nature is absent or decreases when this nature decreases and conversely is the *form* of this nature. The crucial element in this three-stage process is the elimination or exclusion of all accidental characteristics of the nature under investigation. His talk of forms is reminiscent of the Aristotelian substantial forms. Indeed, Bacon's was a view in transition between the Aristotelian and a more modern conception of **laws of nature**. For he also claimed that the form of a nature is the law(s) it obeys. Bacon did favour active experimentation and showed great respect for alchemists because they had had laboratories. In his instance of fingerpost, he claimed that an essential instance of the interpretation of nature consists in devising a **crucial experiment**. Bacon also spoke against the traditional separation between theoretical and practical knowledge and argued that human knowledge and human power meet in one.

See **Confirmation, Hempel's theory of; Nicod; Scientific method**

Further reading: Bacon (1620); Losee (2001)

Base-rate fallacy: Best introduced by the Harvard Medical School test. A test for the presence of a disease has two outcomes, 'positive' and 'negative' (call them $+$ and $-$). Let a subject (Joan) take the test. Let H be the hypothesis that Joan has the disease and $-H$ the hypothesis that Joan doesn't have the disease. The test is highly reliable: it has zero *false negative* rate. That is, the **likelihood** that the subject tested negative given that she does have the disease is zero (i.e., $prob(-/H) = 0$). The test has a small *false positive* rate: the likelihood that Joan is tested positive though she doesn't have the disease is, say, 5 per cent ($prob(+/-H) = 0.05$). Joan tests positive. What is the **probability** that Joan has the disease given that she tested positive? When this problem was posed to experimental subjects, they tended to answer that the probability that Joan has the disease given that she tested positive was very high – very close to 95 per cent. However, given only information about the likelihoods $prob(+/H)$ and $prob(+/-H)$, the question above – what is the posterior probability $prob(H/+)$? – is indeterminate. There is some crucial information missing: the incidence rate (base-rate) of the disease in the population. If this incidence rate is very low, for example, if only 1 person in 1,000 has the disease, it is very unlikely that Joan has the disease even though she tested positive: $prob(H/+)$

would be very small. For prob(H/+) to be high, it must be the case that the prior probability that Joan has the disease (i.e., prob(H)) is not too small. The lesson that many have drawn from cases such as this is that it is a **fallacy** to ignore the base-rates because it yields wrong results in probabilistic reasoning.

See **Confirmation, error-statistical theory of; Probability, prior**

Further reading: Howson (2000)

Bayes, Thomas (1702–1761): English mathematician and clergyman. His posthumously published *An Essay Towards Solving a Problem in the Doctrine of Chances* (1764), submitted to the *Philosophical Transactions of the Royal Society of London* by Richard Price, contained a proof of what came to be known as **Bayes's theorem**.

Further reading: Earman (1992)

Bayes's theorem: Theorem of the **probability** calculus. Let H be a hypothesis and e the evidence. Bayes's theorem says: prob(H/e) = prob(e/H)prob(H)/prob(e), where prob(e) = prob(e/H)prob(H)+prob(e/−H)prob(−H). The unconditional prob(H) is called the prior probability of the hypothesis, the conditional prob(H/e) is called the posterior probability of the hypothesis *given* the evidence and the prob(e/H) is called the **likelihood** of the evidence given the hypothesis.

See **Bayesianism; Probability, posterior; Probability, prior**

Further reading: Earman (1992); Howson and Urbach (2006)

Bayesianism: Mathematical theory based on the **probability** calculus aiming to provide a general framework in which key concepts such as **rationality, scientific method,**

confirmation, evidential support and inductive **inference** are cast and analysed. It borrows its name from a theorem of probability calculus: **Bayes's Theorem.** In its dominant version, Bayesianism is subjective or personalist because it claims that probabilities express subjective (or personal) **degrees of belief.** It is based on the significant mathematical result – proved by **Ramsey** and, independently, by the Italian statistician Bruno de Finnetti (1906–1985) – that subjective degrees of belief (expressed as fair **betting quotients**) satisfy Kolomogorov's axioms for probability functions. The key idea, known as the Dutch-book theorem, is that unless the degrees of belief that an agent possesses, *at any given time*, satisfy the axioms of the probability calculus, she is subject to a Dutch-book, that is, to a set of synchronic bets such that they are all fair by her own lights, and yet, taken together, make her suffer a net loss come what may. The monetary aspect of the standard renditions of the Dutch-book theorem is just a dramatic device. The thrust of the Dutch-book theorem is that there is a *structural incoherence* in a system of degrees of belief that violates the axioms of the probability calculus. Bayesianism comes in two varieties: synchronic and diachronic. Synchronic Bayesianism takes the view that the demand for probabilistic coherence among one's degrees of belief is a logical demand: in effect, a demand for logical consistency. However, the view that synchronic probabilistic coherence is a canon of rationality cannot be maintained, since it would require a non-question-begging demonstration that any violation of the axioms of the probability calculus is positively irrational. Diachronic Bayesianism places **conditionalisation** on centre-stage. It is supposed to be a canon of rationality that agents should update their degrees of belief by conditionalising on **evidence.** The penalty for not doing this is liability to a Dutch-book strategy: the agent can be offered a set of bets *over time* such that (1) each of them taken individually will seem fair to her at the time when it is offered; but (2) taken collectively, they lead her to suffer a net loss, come what may. As is generally recognised, the penalty is there on a certain condition, namely, that the agent *announces in advance* the method by which she changes her degrees of belief, when new evidence rolls in, *and* that this method is different from conditionalisation. Critics of diachronic Bayesianism point out that there is no general proof of the conditionalisation rule.

See **Coherence, probabilistic; Confirmation, Bayesian theory of; Probability, subjective interpretation of**

Further reading: Earman (1992); Howson and Urbach (2006); Sober (2002)

Belief: Psychological state which captures the not necessarily alethic part of **knowledge**. It is a state with propositional content, often captured by the locution 'subject S believes that—' where a proposition is substituted for the solid line (as in: John believes that electrons have charge). Beliefs can be assessed in terms of their **truth** or falsity and in terms of their being justified (warranted) or not. In particular, a justified true belief constitutes knowledge. But beliefs can be justified (e.g., they may be the product of thorough investigation based on the evidence) even though they may (turn out to) be false. *Qua* psychological states beliefs can be causes and effects. But philosophers have been mostly concerned with their normative appraisal: are they appropriately based on reasons and **evidence**? *Qua* psychological states, beliefs can also be either dispositional or occurrent. They are dispositional if their possession is manifested under certain circumstances (e.g., I have the belief that snow is white because I have a **disposition** to assent to the proposition that snow is white). Dispositional beliefs can be possessed without being currently assented to. Beliefs are occurrent when they require current assent – that is, when they are manifested. **Popper** and his followers have argued that science is not about belief and have advanced an epistemology that dispenses with belief altogether. But it is hard to see how the concept of knowledge can be had without the concept of belief. Many philosophers of science (especially followers of **Bayesianism**) have focused on how beliefs change over time.

See **Coherentism; Degree of belief; Foundationalism; Justification; Reliabilism**

Further reading: Williams (2001)

Berkeley, George (1685–1753): Irish philosopher and bishop of the Anglican Church, one of the three most famous eighteenth-century British Empiricists. His basic works are: *A Treatise Concerning the Principles of Human Knowledge* (1710), *Three Dialogues Between Hylas and Philonous* (1713) and *De Motu* (1721). He was an immaterialist in that he denied the existence of matter in so far as 'matter' meant something over and above the collection of perceptible qualities of bodies (ideas). He took issue with the philosophical understanding of matter as an unthinking corporeal substance, a substratum, on which all perceptible qualities of bodies inhere. Berkeley denied the distinction between primary and secondary qualities and argued that all sensible qualities are secondary: they depend on perceiving minds for their existence. He also denied the existence of abstract ideas, that is of abstract forms or **universals,** wherein all particular objects of a certain kind were supposed to partake. Being an empiricist, he thought that all ideas are concrete, and that general ideas (like the idea of triangle) are signs that stand for any particular and concrete idea (for instance, any concrete triangle). Berkeley is considered the founder of **idealism**. He enunciated the principles 'esse' is 'percipi' (to be is to perceive); hence he tied existence to perceiving and to being perceived. It follows that nothing can exist unperceived. Even if there are objects that some (human) mind might not perceive right now, they are always perceived by God. He denied that there is any **causation** in nature, since ideas are essentially passive and inert. He took God to be the cause of all ideas. He explained the fact that there are patterns among ideas (e.g, that fire produces heat), or that some ideas are involuntary (e.g., that when I open my eyes in daylight I see light) by arguing that God has instituted **laws of nature** that govern the succession of ideas. These laws, he thought, do not establish any **necessary connections** among ideas, but constitute regular associations among them. Berkeley has been taken to favour **instrumentalism**. This is true to the extent that he thought that science should not look for causes but for the subsumption of the phenomena under mathematically expressed regularities.

See **Abstraction; Empiricism**

Further reading: Berkeley (1977); Winkler (1989)

Betting quotient: A bet on an outcome P is an arrangement in which the bettor wins a sum S if P obtains and loses a sum Q if P does not obtain. The betting quotient is the ratio $Q/(S+Q)$, where the sum $S+Q$ is the stake and Q/S are the odds. A bet is fair if the agent is indifferent with respect to both sides of the bet, that is, if she does not perceive any advantage in acting as bettor or bookie. The betting quotient is a measure of the agent's subjective degree of belief that P will obtain. According to the Dutch-book theorem, bettors should have betting quotients (and hence subjective **degrees of belief**) that satisfy the axioms of the **probability** calculus.

See **Bayesianism**

Further reading: Howson and Urbach (2006)

Bohr, Niels Henrik David (1885–1962): Danish physicist, one of the founders of modern quantum mechanics. He devised a non-classical **model** of the atom, according to which electrons exist in discrete states of definite energy and 'jump' from one energy state to another. This model solved the problem of the stability of atoms. Bohr initiated the so-called Copenhagen interpretation of quantum mechanics, which became the orthodox interpretation. One of his main ideas was the principle of complementarity, which he applied to the wave-particle duality as well as the classical world and the quantum world as a whole. According to this principle some concepts, or perspectives, or theories, are complementary rather than contradictory in that, though they are mutually exclusive, they are applicable to different aspects of the phenomena. Hence, though they cannot be applied simultaneously, they are indispensable for a full characterisation or understanding of the phenomena. Against **Einstein**, Bohr argued that it does not make sense to think of a quantum object as having determinate **properties** between measurements. The attribution of properties to quantum objects was taken to be meaningful only relative to a choice of a measuring apparatus. He also gave an ontological gloss to Werner Heisenberg's (1901–1976) uncertainty principle, according to which the quantum state offers a complete description of this system and the uncertainty that there is in measuring a property of a system (e.g., its momentum) is not a matter of ignorance but rather a matter of the indeterminacy of the system. Bohr has been taken to favour an instrumentalist construal of scientific theories.

See **Instrumentalism; Quantum mechanics, interpretations of**

Further reading: Murdoch (1987)

Boltzmann, Ludwig (1844–1906): Austrian physicist, the founder of statistical mechanics, which brought thermodynamics within the fold of classical mechanics. In 1903 he succeeded **Mach** as Professor of the Philosophy of Inductive Science, in the University of Vienna. He was a defender of the atomic theory of matter (to which he made substantial contributions) against energetics, a rival theory that aimed to do away with atoms and **unobservable entities** in general. One of his most important claims was that the second law of thermodynamics (the law of increase of entropy) was statistical rather than deterministic. He developed a view of theories according to which theories are mental images that have only a partial similarity with **reality**.

Further reading: de Regt (2005)

Bootstrapping: Theory of **confirmation** introduced by **Glymour**. It was meant to be an improvement over **Hempel**'s positive-instance account, especially when it comes to showing how theoretical hypotheses are confirmed. It takes confirmation to be a three-place relation: the evidence *e* confirms a hypothesis *H relative to* a theory *T* (which may be the very theory in which the hypothesis under test belongs). Confirmation of a hypothesis *H* is taken to consist in the deduction of an instance of the hypothesis *H* under test from premises which include the data *e* and (other) theoretical hypotheses of the theory *T* (where the deduction is such that it is not guaranteed that an instance of *H* would be deduced irrespective of what the data might have been). Though relative to a theory, the confirmation of the hypothesis is *absolute* in that the evidence either does or does not confirm it. The idea of bootstrapping is meant to suggest how some parts of a theory can be used in specifying how the evidence bears on some other parts of the theory without this procedure creating a vicious circle. Glymour's account gave a prominent role to **explanation**, but failed to show how the confirmation of a hypothesis can give scientists reasons to believe in the hypothesis. The objection is that unless probabilities are introduced into a theory of confirmation, there is no connection between confirmation and reasons for belief.

See **Confirmation, Bayesian theory of; Confirmation, Hempel's theory of**

Further reading: Glymour (1980)

Boyd, Richard (born 1942): American philosopher, author of a number of influential articles in defence of **scientific realism**. He placed the defence of realism firmly within a naturalistic perspective and advanced the explanationist defence of realism, according to which realism should be accepted on the grounds that it offers the best **explanation** of the successes of **scientific theories**. He has been a critic of **empiricism** and of **social constructivism** and has claimed that scientific realism is best defended within the framework of a non-Humean metaphysics and a robust account of **causation**.

Further reading: Boyd (1981)

Boyle, Robert (1627–1691): English scientist, one of the most prominent figures of seventeenth-century England. He articulated the **mechanical philosophy**, which he saw as a weapon against Aristotelianism, and engaged in active experimentation to show that the mechanical conception of nature is true. He defended a corpuscular approach to matter. In *About the Excellency and Grounds of the Mechanical Hypothesis* (1674), he outlined his view that all natural phenomena are produced by the mechanical interactions of the parts of matter according to mechanical laws. He also wrote about methodological matters. He favoured consistency, **simplicity**, comprehensiveness and applicability to the phenomena as **theoretical virtues** that theories should possess and argued that his own corpuscularian approach was preferable to Aristotelianism because it possessed these virtues.

Further reading: Boyle (1979)

C

Carnap, Rudolf (1891–1970): German-American philosopher of science, perhaps one of the most important philosophers of science ever. He was a member of the **Vienna Circle** and emigrated to the USA in 1935, where he stayed until his death, holding chairs in the University of Chicago and the University of California, Los Angeles. He made original and substantial contributions to very many areas of the philosophy of science, most notably the structure of **scientific theories**, the logic of **confirmation**, **inductive logic** and semantics. In the 1920s, Carnap's work was focused on epistemological issues and in particular on how the world of science relates to the world of experience. In *The Logical Structure of the World* (1928) he aimed to show how the physical world could emerge from within his constructional system as the inter-subjective point of view, where all physical objects were, in effect, the 'common factors' of the individual subjective points of view. For him the new logic of **Frege** and **Russell** sufficed for the specification and derivation of all relational **concepts**, and since this logic was analytic and **a priori**, he had no place for the Kantian synthetic a priori. He went on to advance a form of **structuralism**, tying content (the material) to subjective experience and making formal **structure** the locus of **objectivity**. His task was to characterise all concepts that may legitimately figure in his system of unified science by means of 'purely structural definite descriptions'. In the 1930s, his attention was shifted to the logic of science, where the latter was taken to be a formal study of the language of science. The key idea, developed in *The Logical Syntax of Language* (1934), was that the development of a general theory of the logical syntax of the logico-mathematical language of science would provide a neutral framework in which: (1) scientific theories are cast and studied; (2) scientific concepts (e.g., *explanation, confirmation, laws* etc.) are explicated; and (3) traditional metaphysical disputes are overcome. The whole project required that a sharp **analytic/synthetic distinction** could be drawn: philosophical statements would be analytic (about the language of science) whereas scientific ones would be synthetic (about the world). His *Testability and Meaning* (1937) marked his turn to issues in semantics. He aimed to liberalise **empiricism**, by weakening the **verifiability** criterion of meaning and replacing it with a criterion based on testability. He developed the technique of **reduction sentences** in an attempt to show how the meaning of theoretical concepts could be specified (if only partially) by reference to tests and experimental situations. In the late 1940s, he plunged into the theory of confirmation, developing a system of inductive logic based on the idea of **partial entailment**. In the 1950s he wrote extensively on the structure of scientific theories and their empirical content. He took an irenic and conciliatory stance on the realism-**instrumentalism** debate. He re-invented the **Ramsey-sentence** approach to scientific theories and tried to show that it holds the key to understanding the semantics of theoretical terms. Throughout his philosophical career, he held on to the analytic-synthetic distinction, though he relativised it to linguistic frameworks. He also held on to a disdain for metaphysics, the proper content of which he took it to concern the choice of linguistic frameworks.

See **Atomism, semantic; Convention; Definition, explicit; Explication; External/internal questions; Formal mode vs material mode; Induction, the problem of; Logical positivism; Principle of tolerance; Probability, logical interpretation of; Protocol sentences; Syntactic view of theories**

Further reading: Carnap (1928, 1936, 1950b, 1974)

Cartwright, Nancy (born 1944): American philosopher of science, author of *How the Laws of Physics Lie* (1983) and *The Dappled World* (1999). She argued that, based on inference to the most likely cause, **empiricism** can lead to warranted belief in the existence of **unobservable entities**. She favoured **entity realism** and argued against theory realism, especially when it came to high-level and abstract scientific theories. She claimed that the laws of physics lie. Her point concerned mostly the fundamental or abstract laws as well as the **covering-law model** of **explanation**. If laws explain by 'covering' the facts to be explained, the explanation offered will be false. If the laws are hedged by *ceteris paribus* clauses, they become truer but do not 'cover' the facts anymore; hence, they do *not* explain them. Cartwright has taken capacities to be prior to laws. There are laws in nature insofar as there are nomological machines to sustain them, where nomological machines are constituted, at least partly, by stable capacities. Cartwitght's capacities are causal **powers**. The

world, according to Cartwright, is dappled, with pockets of order and disorder and no overall unified structure.
See **Unification**
Further reading: Cartwright (1983, 1999)

Categorical properties see **Dispositions; Powers; Properties**

Causal graphs: Graphical representations of causal structure in terms of relations of probabilistic dependence among variables. A directed acyclic graph (DAG) consists of: a set of nodes corresponding to variables; a set of edges (directed arrows), representing dependencies; a **conditional probability** distribution for each node; the absence of any directed cycles. They are also known as Bayes nets (mostly because of the use of subjective prior probabilities and the reliance on Bayesian **conditionalisation**) and they can be used for causal **inference**, prediction and **explanation**. A Bayes net is characterised by the satisfaction of Markov's condition. Take a node A and call its parents all nodes that have edges into A and its grandparents all nodes that have edges into A's parents. Then, Markov's condition says that the probability of a variable depends only on its parents and not on its grandparents. In other words, Markov's condition is a condition of probabilistic independence. The probability of dying given that one has developed lung cancer *and* one has been smoking is the same as the probability of dying given that one has developed lung cancer. Interventions are easily representable in causal graphs (by breaking the links among variables). The stability of the graph under (actual and counterfactual) interventions represents the stability of causal structures. The theory of causal graphs has been developed (among others) by the computer scientist Judea Pearl (born 1936) and **Glymour**.
See **Causation; Probability**
Further reading: Woodward (2003)

Causal process: As opposed to **events,** which are happenings localised in **space** and **time,** processes are extended in space and time. Examples of processes include a light-wave travelling from the sun, or the movement of a ball. Material objects that persist through time can be seen as processes. In the Special Theory of Relativity, a process is represented by a world-line in a Minkowski diagram, whereas an event is represented by a point. A causal process is characterised by causal unity, for example, the persistence of a quality or the possession of some characteristic. According to **Salmon,** causal processes are the fundamental elements of the mechanistic approach to causation: they constitute the **mechanisms** that link cause and effect and transmit causal influence. Causal processes, argues Salmon, are those processes that are capable of transmitting a mark. A mark is a modification of the structure of the process by means of a single local interaction. The idea of marking a causal process goes back to **Reichenbach. Russell**'s causal lines are similar to causal processes. A causal line captures the persistence of some characteristic in a process, for example, the constancy of quality, or the constancy of structure.

　See **Causation**
　Further reading: Dowe (2000); Salmon (1984)

Causal relata: Those that are related by a causal relation, namely, the cause and the effect. According to the standard approach, the causal relata are **events** (like hittings, smashings, pushings, throwings etc.). To say that c caused e is to say that there exist unique events c and e, and c caused e. For instance, to say that the short circuit caused the fire is to say that there exist unique events c, such that c is a short circuit, and e, such that e is a fire, and c caused e. Some philosophers (notably **Mellor**) take facts to be the causal relata, where facts may be seen as whatever true propositions express. The claim is that to say that *c* caused *e* is to say that the fact that *c* caused the fact *e* (e.g., the fact that John fell from the stairs caused the fact that he broke his leg). While events are concrete and occur in particular spatio-temporal locations, for example, the sinking of the Titanic, facts (e.g., the fact that the Titinic sank) are usually taken to be abstract – with no particular spatio-temporal location.

See **Causation**

Further reading: Sosa and Tooley (1993)

Causal theory of reference: Introduced by **Kripke**, it identifies the semantic value (denotation/reference) of a name with the individual/entity that this name refers to. According to it, the reference of a proper name is fixed by a causal-historical chain, which links the current use of the name with an introducing event, which associated the name with its bearer. Descriptions associated with the name might be false, and yet the name-users still refer to the named individual, insofar as their use of the name is part of a causal transmission-chain which goes back to the introducing event. The thrust of the causal theory is that the relation between a word and an object is *direct* – a direct causal link – unmediated by a **concept**. In particular, the causal theory dispenses with sense as a reference-fixing device. The theory was extended to cover the reference of natural-kind and physical-magnitude terms, mostly by **Putnam**. The reference of a natural-kind term is fixed during an introducing event, that is, an event during which the term is attached to a substance, or a kind, where samples of this substance or instances of this kind are present and ground the term. More generally, when confronted with some observable phenomena, it is assumed that there is a physical entity that causes them. Then we (or indeed, the first person to notice them) dub this entity with a term and associate this entity with the production of these phenomena. The reference of the term has been fixed *existentially* as the entity causally responsible for certain effects. The chief attraction of the causal theory is that it lends credence to the claim that even though past scientists had partially or fully incorrect beliefs about the properties of a causal agent, their investigations were continuous with the investigations of subsequent scientists, since their common aim has been to identify the nature of the same causal agent. There is a sense in which the causal theory of reference makes reference-continuity in theory-change all too easy. If the reference of theoretical terms is fixed purely existentially, insofar as *there is* a causal agent behind the relevant phenomena, the term is bound to end up referring to it.

See **Description theories of reference; Natural kinds; Sense and reference**

Further reading: Devitt and Sterelny (1987); Kripke (1980); Unger (1983)

Causation: The relation between cause and effect. What is the nature of the connection between cause and effect: how and in virtue of what is the cause related to the effect? There have been two broad approaches to this issue: causation as a relation of *dependence* and causation as a relation of *production*. On the dependence approach, causation is a kind of robust relation between discrete events: to say that *c* causes *e* is to say that *e* suitably depends on *c*. On the production approach, to say that *c* causes *e* is to say that something *in* the cause produces (brings about) the effect or that there is something (e.g., a **mechanism**) that links the cause and the effect. There have been different ways to cash out the relation of dependence: nomological dependence (cause and effect fall under a law); counterfactual dependence (if the cause hadn't happened, the effect wouldn't have happened); probabilistic dependence (the cause raises the probability of the effect). Similarly, there have been different ways to cash out the concept of production, but the most prominent among them are cast in terms of something being transferred from the cause to the effect (e.g., a property, or some physical quantity – force, energy etc.). A key thought in the production approach is that cause and effect are connected by means of a local mechanism.

Nomological dependence: On this view that goes back to **Hume,** causation reduces to a relation of spatiotemporal contiguity, succession and **constant conjunction** (regularity) between distinct events. That is, *c* causes *e* iff

1. *c* is spatiotemporally contiguous to *e*;
2. *e* succeeds *c* in time; and
3. all **events** of type *C* (i.e., events that are like *c*) are regularly followed by (or are constantly conjoined with) events of type *E* (i.e., events like *e*).

A corollary of this view is that there is no **necessary connection** between the cause *c* and the effect *e*. Some Humeans (most notably **Mill** and Mackie), advanced more sophisticated versions of the Regularity View of Causation. A prominent thought has been that causation should be analysed in terms of *sufficient and necessary conditions* (roughly, an event *c* causes an event *e* iff there are event-types *C* and *E* such that *C* is necessary and sufficient for *E*). Another one has been that to call an event *c* the cause of an event *e* is to say that there are event-types *C* and *E* such that *C* is an *insufficient* but *necessary* part of an *unnecessary* but *sufficient* condition for *E* – also known as **inus-conditions.** In all versions of the Regularity View, whether or not a sequence of events is causal depends on things that happen elsewhere and at other times in the universe, and in particular on whether or not this particular sequence instantiates a regularity. Yet, it seems there can be causation without regularity. This is the case of singular causation. Conversely, there can be regularity without causation. There are cases in which events regularly follow each other (like the night always follows the day) without being the cause of each other. Humeans have been inegalitarians towards regularities. They have tried to characterise the kind of regularity that can underpin causal relations by tying causation to **laws of nature**.

Counterfactual dependence: Causation is defined in terms of the *counterfactual dependence* of the effect on the cause: the cause is counterfactually necessary for the effect. For instance, to say that the short circuit caused the fire is to say that *if the short circuit hadn't happened, the fire wouldn't have ensued*. More precisely, **Lewis** defined causation by reference to a *causal chain* of counterfactually dependent events, where a sequence of events $< c, e, e' \ldots >$ is a chain of counterfactual dependence if and only if e counterfactually depends on c, e' counterfactually depends on e and so on. This move is meant to enforce that causation is a transitive relation among events (i.e., if c causes e and e causes e', then c causes e'). So one event is a cause of another if and only if there exists a causal chain leading from the first to second. Lewis articulated a rather complicated logic of **counterfactual conditionals**. Though Lewis's theory is meant to capture singular causation, regularities do enter the counterfactual approach in a roundabout way: as means to capture the conditions under which counterfactual assertions are true. Problems for the counterfactual theory occur in cases of causal **overdetermination** and pre-emption.

Probabilistic dependence: Causes raise the *chances* of their effects, namely, the probability that a certain event happens is higher if we take into account its cause than if we don't. In broad outline, c causes e iff (1) the **probability** of e given c is greater than the probability of e given not-c and (2) there is no other factor c' such that the probability of e given c and c' is equal to the probability of e given not-c and c'. This latter condition is called screening off. Theories of probabilistic causation rely on the claim that there can be causation even when there is no regularity or (deterministic) laws. A problem faced by all theories of probabilistic causation is that there are circumstances in which c causes e while lowering the probability that it will happen.

Manipulation theories: Causes are recipes for producing or preventing their effects. This thought is normally cast in terms of *manipulability*: causes can be manipulated to bring about certain effects. **Von Wright** developed this view into a full-blown theory of causation. He took it that what confers on a sequence of events the character of causal connection is the possibility of subjecting causes to experimental test by interfering with the 'natural' course of events. Since manipulation is a distinctively human action, he concluded that the causal relation is dependent upon the concept of human action. But his views were taken to be too anthropomorphic. Recently, there have been important attempts to give a more objective gloss to the idea of manipulation. James Woodward (born 1946) introduced a notion of **intervention** that is not restricted to human action and argued that a relationship among some magnitudes X and Y is causal if, were one to intervene to change the value of X appropriately, the relationship between X and Y would remain invariant *but* the value of Y would change, as a result of the intervention on X.

Transference models: The idea that causation is a productive relation goes back at least to **Descartes** who put forward the *transference model* of causation: when x causes y a property of x is communicated to y. He thought that this view is an obvious consequence of the principle 'Nothing comes from nothing'. Recently, transference models have been tied to physical **properties** such as energy-momentum. They have given rise to mechanistic theories of causation according to which there is a mechanism that connects cause and effect. According to **Salmon**'s *mechanistic* approach, an event c causes an event e if and only if there is a **causal process** that connects c and e. Later on, Salmon took causation to consist in the exchange or transfer of some *conserved quantity*, such as energy-momentum or charge. But according to Phil Dowe a process is causal if and only if it *possesses* a conserved quantity. Though this theory seems plausible when it comes to physical causation, it is questionable whether it can be generalised to cover all cases of causation and especially the cases in the special sciences (economics, psychology etc.).

See **Causation, singular; Explanation; Laws of nature**

Further reading: Dowe (2000); Eells (1991); Lewis (1973a); Mackie (1974); Psillos (2002); Sosa and Tooley (1993); Woodward (2003)

Causation, direction of: Causes precede their effects in time. Why is this so? Some philosophers (including **Hume**) thought this feature is conceptually constitutive of **causation**: the direction of causation is the direction of time. So, there cannot be 'backward causation', namely, causal relations in which the effect precedes in time the cause. Others think that the direction of causation cannot be settled a priori. Even if, in the actual world, the causal order has a preferred, forward-looking, direction, in other possible worlds this direction might be reversed. The relation between causal order and temporal order is a matter of controversy. Many philosophers try to define the direction of causation independently of the concept of time, so that they can then explain the direction of **time** in terms of the direction of causation. **Reichenbach,** for instance, explained the temporal order in terms of the direction of causal relations, which he understood in terms of the asymmetry exhibited by statistical forks that capture common-cause structures, namely, that common causes screen off their effects and not conversely. **Lewis** took the direction of causation to be a function of the asymmetry of **counterfactual dependence** and, in particular, of the fact that the past is counterfactually independent of the present, since it would remain the same whatever we did now, while the future would not. Others take it that the direction of causation is determined in the actual world by some contingent physical principle, for example, the second law of thermodynamics.

Further reading: Horwich (1987); Price (1996)

Causation, singular: According to many non-Humeans, **causation** is essentially singular: a matter of *this* causing *that*. John Curt Ducasse (1881–1969) argued that what makes a sequence of events causal is something that happens there and then: a local tie between the cause and the effect, or an intrinsic feature of the particular sequence. Ducasse's single-difference account, roughly that an event *c* causes an event *e* if and only if *c* was the last – or, the only – difference in *e*'s environment before *e* occurred, takes causation to link individual events independently of any regular association that there may or may not exist between events like the cause and events like the effect.

Further reading: Ducasse (1969)

Certainty: The requirement that in order for a **belief** to be warranted it must be such that it is impossible that this belief be false. Though there is a sense in which certainty is a psychological state, in traditional epistemology it has been taken to be a quasi-logical requirement of indubitability: a belief is certain if it is impossible to be doubted. In the Cartesian tradition, **knowledge** has been equated with certainty. It has been suggested that a subject is in a state of knowledge if and only if the subject's relevant beliefs have been produced by processes/methods that infallibly yield true beliefs. But certainty is not an uncontroversial **explication** of our pre-analytical concept of knowledge. Rather, it is part of a highly contentious reconstruction of the concept of knowledge, which is based on an illegitimate transfer of features of mathematical knowledge to knowledge in general. According to epistemological **naturalism**, knowledge does not require certainty; reliable belief-forming processes are enough to yield knowledge.

See **Coherentism; Foundationalism; Reliabilism; Scepticism**

Further reading: Klein (1984)

Ceteris paribus **laws:** Laws that hold under certain conditions, when other things are equal (or normal). The *ceteris paribus* clause is supposed to hedge the universal applicability (and exceptionless character) of the law. For instance, *ceteris paribus*, if demand for a given product exceeds supply, prices will rise. Here, it is obvious that the *ceteris paribus* clause is meant to ground the possibility of exceptions: the law holds as long as all other factors (e.g., the existence of an alternative product) remain constant. There seems to be agreement that if there are laws at all in the so-called special (non-fundamental) sciences (like psychology or economics), these are *ceteris paribus* laws. But there is disagreement when it comes to fundamental physics. Here, there are philosophers (notably **Earman**) who take the laws of fundamental physics to be strict, while others (notably **Cartwright**) take them too to be *ceteris paribus*. A popular claim is that *ceteris paribus* laws are vacuous, since they assert that things are thus-and-so, unless they are otherwise! *Ceteris paribus* laws forfeit deducibility from strict physical laws; hence, holding on to *ceteris paribus* laws is a way to deny the reducibility of a domain to another. For others, however, the existence of *ceteris paribus* laws in a domain amounts to the admission that the science that covers this domain is not yet mature.

See **Laws of nature; Reduction**

Further reading: Lange (2000)

Chance: The objective (single-case) **probability** of an event to happen. According to the deniers of **determinism,** chances (i.e., objective probabilities other than one and zero) are objective **properties** of the world. Advocates of the relative frequency interpretation of probability accommodate chances by taking them to be limiting relative frequencies.

See **Principal principle; Probability, frequency interpretation of; Propensity**

Further reading: Albert (2000); Sklar (1995)

Coherence, probabilistic: The property of systems of **degrees of belief** in virtue of which no Dutch-book can be made against it. A set of degrees of beliefs is coherent if and only if they satisfy the axioms of **probability**. This last claim is known as the Ramsey–de Finetti theorem.

See **Bayesianism; Conditionalisation**

Further reading: Howson and Urbach (2006)

Coherentism: Holist and non-linear theory of **justification**. It denies that there is any division between beliefs into basic and derived. All beliefs within a system are justified insofar as the system as a whole is justified. Hence, justification accrues to the system of beliefs as a whole and not (primarily) to the individual beliefs that constitute it. The justification of a system of beliefs is a function of its coherence. Coherence cannot just be logical consistency, since the latter is too weak a condition. Any logical coherent system of beliefs – no matter how problematic – would be justified. Besides, all coherent systems of beliefs would be equally justified. Advocates of coherentism favour explanatory coherence: each **belief** of the system should either explain some other beliefs or be explained by other beliefs. The demand of coherence can then be seen as a demand for explanatory **unification**. The basic intuition behind coherentism is the thought (made popular by **Neurath** and Donald Davidson (1917–2003) that only a belief can justify a belief. But then how does a coherent system of beliefs relate to the world? Coherentism cannot easily explain the friction between the system of the beliefs and the world unless it gives some special epistemic status to some beliefs (e.g., observational beliefs) which are not justified exclusively on the basis of their internal or inferential connections with other beliefs. This kind of modified coherentism has been defended by **Quine** in his image of the web of beliefs.

See **Foundationalism**

Further reading: BonJour (1985); Williams (2001)

Concept empiricism: The view that all meaningful **concepts** must get their meaning from experience. Hence, the meaning of a concept must be either directly given in experience (e.g., by means of ostension) or specified by virtue of the meanings of concepts whose own meaning is directly given in experience. Traditionally, it has been associated with the rejection of innate ideas. Though there have been very austere versions of concept empiricism according to which even the concepts of logic and mathematics get their meaning from experience, concept empiricism is consistent with the view that some concepts get their meaning independently of experience by means of stipulations. Hence, concept empiricism is compatible with the **analytic/synthetic distinction**. Concept empiricism might be plausible when it comes to concepts that refer to whatever is immediately given in experience (e.g., colour concepts). A standard objection, however, is that some concepts must be innate since acquiring concepts from experience requires the application of some concepts. For instance, to acquire the concept 'red' from experience it is required that one is able to discriminate reliably between red things and, say, green things – but this presupposes a concept of similarity. Some concept empiricists allow for the existence of innate **mechanisms** of learning concepts (e.g., an innate mechanism which sorts things out in terms of their similarity to each other), though deny that the concepts themselves are innate. In any case, concept empiricism faces the problem of specifying the meaning of theoretical concepts (e.g., the concepts of **scientific theories**). Traditionally, empiricists thought that the meaning of theoretical concepts is fixed by explicit definitions. Arguments in favour of semantic holism suggested that the meaning of theoretical terms is fixed in a holistic way in virtue of the relations to other concepts and the theories in which they are embedded.

See **Carnap; Definition; Empiricism; Holism, semantic; Judgement empiricism; Mill**

Further reading: Reichenbach (1951); Russell (1912); Sellars (1963)

Concepts: The constituents of thoughts; alternatively, the content of words: what words mean. There have been two broad views about what concepts are. Both take concepts to be *of* something (e.g., the concept of a horse or the concept of triangularity). The first view takes the possession of concepts to be an ability, namely, an ability to think thoughts about something. Hence, to possess the concept of *X* is to be able to think thoughts that something is (or falls under) *X*. According to the second view, to possess a concept is to stand in a relation to a certain entity. Hence, concepts are entities, which can nonetheless be about something. In its most popular version, this second view takes concepts to be **abstract entities** (intensions, meanings). This is taken to guarantee the objective character of thought. The existence of concepts is taken to be independent of minds and thoughts acquire their content by being related to (or being about) concepts.

See **Frege**

Further reading: Fodor (1998)

Condition, necessary: A condition (factor) *F* is said to be necessary for another condition (factor) *G* if, when *F* does not obtain, *G* does not obtain either. In logic, *F* is necessary for *G* if *not-F* implies *not-G* (which is logically equivalent to '*G* implies *F*'). For example, a necessary condition for making apple pie is using apples. And a necessary condition for being human is being animal. A condition can be necessary for something without also being sufficient for it.

See **Condition Sufficient**

Condition, necessary and sufficient see **Condition, necessary; Condition, sufficient**

Condition, sufficient: A condition (factor) *F* is said to be sufficient for another condition (factor) *G* if, when *F* obtains, *G* obtains too. In logic, *F* is sufficient for *G* if *F* implies *G* (which is logically equivalent to '*not-G* implies *not-F*').

For example, a sufficient condition for being coloured is being red. A condition F can be sufficient for G without also being necessary for it.

See **Condition, necessary**

Conditional probability: The **probability** that (event or outcome) q will happen, on the condition (or given) that p has happened. The (unconditional) probability of q may or may not be the same as the conditional probability of q given p. For instance, the probability of drawing an ace of spades from a deck of cards is $\frac{1}{52}$, but the probability of drawing an ace of spades given that the card drawn is black is $\frac{1}{26}$. Two events (or outcomes or propositions) q and p are *independent* of each other if and only if the conditional probability of q given p is equal to the (unconditional) probability of q.

Further reading: Howson and Urbach (2006)

Conditionalisation: The view that agents should update their **degrees of belief** by conditionalising on the **evidence**. The transition from the old probability $\text{Prob}_{old}(H)$ that a hypothesis H has had to a new one $\text{Prob}_{new}(H)$ in the light of incoming evidence e, is governed by the rule: $\text{Prob}_{new}(H) = \text{Prob}_{old}(H/e)$, where e is the total evidence, and $\text{Prob}_{old}(H/e)$ is given by **Bayes's theorem**. Conditionalisation can take two forms depending on whether the **probability** of the learned evidence is unity or on whether only logical truths get probability equal to unity. Conditionalising on the evidence is a *logical* updating of degrees of belief. It's not ampliative. It does not introduce new content; nor does it modify the old one. It just assigns a new probability to an old opinion.

See **Ampliative inference; Bayesianism; Dutch-book; Total evidence, principle of**

Further reading: Earman (1992); Howson and Urbach (2006)

Confirmation: The relation between **evidence** and theory in virtue of which the evidence supports the theory. There are three conceptions of confirmation: *qualitative* confirmation, namely, evidence *e* confirms or supports hypothesis *H*; *comparative* confirmation, namely, evidence *e* confirms hypothesis *H* more strongly than it confirms hypothesis *H'*; and, finally *quantitative* confirmation, namely, the degree of confirmation of hypothesis *H* by evidence *e* is *r*, where *r* is a real number. Current theories of confirmation rely heavily on probabilistic relations between the evidence and the theory.

Further reading: Achinstein (2001); Hempel (1965)

Confirmation, absolute vs relative: A piece of **evidence** *e* absolutely confirms some hypothesis *H* if the probability of *H* given *e* (i.e., prob(H/e) is greater than a fixed number *r*, where *r* should be between $1/2$ and 1. Accordingly, *e* is evidence for *H* only if *e* is not evidence for the negation of *H*. This requirement is meant to capture the view that evidence should provide a *good reason* to believe. Relative **confirmation**, in contrast, is incremental confirmation: a piece of evidence *e* confirms some hypothesis *H* if the probability of *H* given *e* (i.e., prob(H/e) is greater than the probability of *H* in the absence of *e* (i.e., prob(*H/-e*). Accordingly, relative confirmation is a relation of positive relevance, namely, that a piece of evidence confirms a theory if it increases its **probability**, no matter by how little.

Further reading: Achinstein (2001); Carnap (1950)

Confirmation, Bayesian theory of: According to **Bayesianism:** (1) **confirmation** is a relation of positive relevance, namely, a piece of evidence confirms a theory if it increases its **probability**; (2) this relation of confirmation is captured by **Bayes's theorem**; (3) the only factors relevant to confirmation of a theory are its prior probability, the

likelihood of the **evidence** given the theory and the probability of the evidence; (4) the specification of the prior probability of (prior **degree of belief** in) a hypothesis is a purely subjective matter; (5) the only (logical-rational) constraint on an assignment of prior probabilities to several hypotheses should be that they obey the axioms of the probability calculus; (6) hence, the reasonableness of a belief does not depend on its content; nor, ultimately, on whether the belief is made reasonable by the evidence; and (7) degrees of belief are probabilities and belief is always a matter of degree. This approach has had many successes. Old problems, like the **paradox of the ravens** or the **grue** problem, have been resolved. New issues, like the **problem of old evidence**, have been, to some extent, resolved. But there is still a pervasive dissatisfaction with subjective Bayesianism. This dissatisfaction concerns all of (1) to (6) above, but is centred mostly around the point that subjective Bayesianism is *too* subjective to offer an adequate theory of confirmation and of rational **belief**.

See **Confirmation, absolute vs relative; Probability, subjective interpretation of**

Further reading: Achinstein (2001); Carnap (1950); Howson and Urbach (2006)

Confirmation, error-statistical theory of: Developed by Deborah Mayo, it rests on the standard Neyman–Pearson statistics and utilises error-probabilities understood as objective frequencies. Error-probabilities do not characterise the degree of **confirmation** of a hypothesis. They refer to the experimental process itself and specify how reliably it can discriminate between alternative hypotheses. The error-statistical approach is connected with severe testing of experimental hypotheses using reliable test procedures, that is, test procedures that have a high probability to detect an error if there is one and not to register an error if there is none. Critics of this approach, mostly advocates of **Bayesianism**, argue that it commits the **base-rate fallacy**, since it forfeits assigning prior probabilities to the hypotheses under test.

See **Probability; Statistical testing**

Further reading: Mayo (1996)

Confirmation, Hempel's theory of: It is based on three conditions. (1) Entailment Condition (EC): if e entails H, then e confirms H. (2) Consistency Condition (CC): if e confirms H, and e confirms H^* then H and H^* are logically consistent. (3) Special Consequence Condition (SCC): if e confirms H, and H entails H', then e confirms H'. There is a fourth condition that is intuitively plausible. (4) Converse Consequence Condition (CCC): if e confirms H, and H'' entails H, then e confirms H''. But the first three conditions conjoined with the fourth lead to paradox: a piece of **evidence** can confirm any hypothesis whatever. The proof is this. Take e to be that Mars describes an ellipse. Take H to be that the moon is made of green cheese. e entails itself, and hence, by (1), e confirms itself. Since e confirms e and $(e \ \& \ H)$ entails e, by (4), e confirms $(e \ \& \ H)$. Since $(e$ and $H)$ entails H, by (3), e confirms H. So that Mars describes an ellipse confirms the hypothesis that the moon is made of green cheese! Condition (4) leads to the **tacking paradox. Hempel** jettisoned (4) and endorsed the first three conditions. His own account of **confirmation**, known as 'instance confirmation', is based on a modification of **Nicod**'s criterion. Take the hypothesis H: *All As are B*, that is *for all x if Ax then Bx*. The *development* of H with respect to the data already available is the conjunction of all its positive instances in the data. For instance, if H has three positive instances among individuals a_1, a_2 and a_3 (i.e., $Aa_1 \ \& \ Ba_1$, $Aa_2 \ \& \ Ba_2$, $Aa_3 \ \& \ Ba_3$), the development *dev(H)* of H is: (if Aa_1

then Ba_1) & (if Aa_2 then Ba_2) & (if Aa_3 then Ba_3). Then, some piece of evidence e directly confirms H just in case e entails *dev(H)* for the class of individuals mentioned in e. And some piece of evidence e confirms H just in case e directly confirms every member of a set of sentences K such that K entails H. Hempel's theory falls foul of the **paradox of the ravens** of the **grue** problem.

See **Bootstrapping**

Further reading: Glymour (1980); Hempel (1965)

Conjectures and refutations: Methodology of science defended by **Popper**. According to his tetrad: (1) scientists stumble over some empirical problem; (2) a theory is proposed (conjectured) as an attempt to solve the problem ('tentative solution'); (3) the theory is tested by attempted refutations ('error elimination'); (4) if the theory is refuted, a new theory is conjectured in order to solve the new problem; if the theory is corroborated, it is tentatively accepted.

See **Corroboration; critical rationalism**

Further reading: Popper (1963)

Consilience of inductions: see **Whewell**

Constant conjunction: Expression introduced by **Hume** to refer to the association (co-occurrence) of events of type C and events of type E, for example, two billiard balls hitting each other and flying apart. The observation of a constant conjunction of Cs and Es conditions the mind, according to Hume, to form the **belief** that upon the fresh perception of a C, an E will (or *must*) follow. For Hume, it is this constant conjunction that is involved in our deeming causal a sequence of events.

See **Causation; Necessary connection**

Further reading: Hume (1739); Psillos (2002)

Constructive empiricism: View about science according to which (1) science aims at empirically adequate theories and (2) **acceptance** of scientific theories involves belief only in their **empirical adequacy** (though acceptance involves *more* than belief, namely, commitment to a theory). It has been introduced and defended by **van Fraassen**, which he took it to be a view of science suitable for empiricists. Constructive empiricism differs from **logical positivism** in many ways, but a central difference is that it does not split the language of science into two mutually disjoint sets: the set of **theoretical terms** and predicates and the set of observational terms and predicates. Yet, constructive empiricism is tied to a sharp distinction between observable and **unobservable entities**. Drawing the distinction in terms of entities allows for the description of observable entities to be fully theory laden. However, even theoretically described, an entity does not cease to be observable if a suitably placed observer *could* perceive it with the naked eye. Observability is offered a privileged epistemic role within constructive empiricism: it sets the limits of what is epistemically accessible. This move presupposes that there is a natural and non-arbitrary way to draw a line between observable and unobservable *entities* – a claim that can be thoroughly contested.

See **Empiricism; Scientific realism**

Further reading: Ladyman (2002); Psillos (1999); van Fraassen (1980)

Context of discovery vs context of justification: Distinction introduced by **Reichenbach** to mark the difference between the processes by which **scientific theories** are invented and the logical and methodological procedures by which they are appraised after they have been formulated. The context of discovery was supposed to belong to psychology while the context of **justification** was supposed to be the proper domain of philosophy of science. Developing philosophical theories of science amounted to devising a rational pattern for science, a pattern in which scientific methodology for theory-testing, **explanation** and theory-acceptance have a fixed and precise logical form.

See **Inductive logic**

Further reading: Reichenbach (1951)

Convention: Generally, a statement held true by means of decision. **Poincaré** argued that the principles of geometry and of mechanics were conventions. For him, conventions are general principles which are held true, but whose **truth** can neither be the product of **a priori** reasoning nor be established by a posteriori investigation. The choice of conventions, Poincaré thought, was not arbitrary since some principles were more convenient than others: considerations of **simplicity** and unity, as well as certain experiential facts, should 'guide' the choice of conventional principles. The logical positivists extended **conventionalism** to logic and mathematics, arguing that the only distinction there is is between, on the one hand, empirical (synthetic a posteriori) principles and, on the other hand, conventional (analytic a priori) ones. **Quine** offered a deep challenge to the view that logic was a matter of convention. Logical truths, he argued, cannot be true by convention simply because, if logic were to be derived from conventions, logic would be needed to *infer* logic from conventions.

See **Analytic/synthetic distinction; Principle of tolerance**

Further reading: Giedymin (1982); Poincaré (1902)

Conventionalism: The philosophical view that says of a certain kind of truths that they are truths by **convention** and not by dint of any kind of facts. Kinds of conventionalism are: geometrical, mathematical, logical and methodological. Methodological conventionalism is the view (defended among others by **Popper**) that the basic methodological rules or canons are freely chosen conventions almost on a par with the rules of chess. Once, however, these conventions are adopted, they define the game of science and the legitimate moves within it.

See **Grünbaum; Poincaré**

Further reading: Reichenbach (1958)

Convergence of opinion: Technical result showing that the actual values assigned to prior probabilities do not matter much since they 'wash out' in the long run; that is, they converge on the same value. Suppose, for instance, that a number of individuals assign different subjective prior probabilities to some hypothesis H. Suppose further that a sequence of experiments is performed and the individuals update their prior probabilities by means of **conditionalisation**. It can be proved that, after a point, their posterior probabilities converge on the same value. This result is taken to mitigate the excessive subjectivity of **Bayesianism**.

See **Probability, prior; Probability, subjective interpretation of**

Further reading: Earman (1992); Howson and Urbach (2006)

Copenhagen interpretation of quantum mechanics see **Bohr; Quantum mechanics, interpretations of**

Copernicus, Nicolaus: (1473–1543): Polish astronomer, author of the posthumously published *On the Revolutions of the Celestial Spheres*, which defended a heliocentric model of the solar system. Before Copernicus, the dominant astronomical theory was Claudius Ptolemy's (c. 85–c. 165). He had assumed that the earth is immobile and at the centre of the universe and that all planets (including the moon and the sun) were revolving in circular motions around the earth. To save the appearances of planetary motion, Ptolemy had devised a system of deferents and epicycles. Copernicus's heliocentric model did use circular motions and epicycles, though it made the earth move around the sun. The unsigned preface of *On the Revolutions* placed the content of the book firmly within the save-the-phenomena astronomical tradition. As it turned out, the preface was written by Andreas Osiander, a Lutheran theologian. For Copernicus himself the theory he put forward in the book was *true*. He based this claim mostly on considerations of harmony and **simplicity**. Copernicus was influenced by neo-platonism, a doctrine that motivated a number of thinkers in the late Middle Ages. Its central claim was that nature was fundamentally mathematical and hence that it exhibits a mathematical harmony.

Further reading: Kuhn (1957)

Correspondence rules: *Mixed* sentences that link **theoretical terms** with observational ones. A typical example of a correspondence rule is this: the theoretical term 'mass' is connected with the observational predicate 'is heavier than' by means of the rule C: 'the mass of body **u** is greater than the mass of body **v** if **u** is heavier than **v**'. They played a central role in the logical positivists' account of theories as formal axiomatic systems. It was a central thought of the logical positivists that a scientific theory need not be completely interpreted to be meaningful and applicable. They claimed that it is enough that only *some*, the so-called 'observational', terms and predicates be interpreted: those whose application is based solely on naked-eye observations. Correspondence rules were supposed to confer partial interpretation on theoretical terms. However, the correspondence rules muddle the distinction between the analytic (meaning-related) and synthetic (fact-stating) part of a scientific theory, which was central in **logical positivism**. For, on the one hand, they specify (even partly) the meaning of theoretical terms and, on the other, they contribute to the factual content of the theory.

See **Analytic/synthetic distinction; Semantic view of theories; Syntactic view of theories; terms, observational and theoretical**

Further reading: Carnap (1956); Suppe (1977)

Corroboration: Technical term introduced by **Popper** to distinguish his view from those inductivists who think that evidence can confirm a hypothesis. Corroborated is a hypothesis that (1) has not yet been refuted and (2) has stood up to severe tests (i.e., attempts at refutation). For Popper, hypotheses are never confirmed by the evidence. If observations do not falsify a hypothesis, the hypothesis does not become probable. It becomes corroborated. But the concept of corroboration cannot explain why it is rational for scientists to base their future predictions on the best corroborated theory. To do this, it is inevitable for them to accept some kind of **principle of induction**. Corroboration should not be confused with **verisimilitude**: from the fact that a theory T is more corroborated than theory T' it does not follow that T is closer to the **truth** than T'.

See **Confirmation; Conjectures and refutations**

Further reading: Popper (1959)

Counterfactual conditionals: Conditionals of the form: if p hadn't happened, then q wouldn't have happened, or if p had happened, then q would have happened. Typically, they are symbolised thus: $p \,\square\!\!\rightarrow q$ or $-p \,\square\!\!\rightarrow -q$. The truth-conditions of such conditionals cannot be specified by means of truth-tables (and in particular by the truth-table of the material implication), because their antecedent is false (given that p did or did not actually happen), and hence the counterfactual conditional would end up being trivially true. Given the centrality of counterfactual conditionals in the understanding and analysis of a number of philosophical concepts (e.g., **causation, dispositions** and others), there have been a number of attempts to specify their logic. **Goodman** suggested that a counterfactual conditional $p \,\square\!\rightarrow q$ is true if and only if its antecedent p nomologically implies, given certain other prevailing conditions, the **truth** of its consequent q. Take, for instance, the counterfactual that if this match had been struck, it would have lit. On Goodman's theory, this statement is true because the antecedent (the match is struck), together with other certain facts (e.g., that the match is dry, there is oxygen present etc.) *and* the laws of nature, imply the consequent (the match lights). The conditions under which a counterfactual is true are linked to the presence of laws, which determine that, given the antecedent, the consequent must obtain. An alternative approach, due mainly to **Lewis**, uses the notion of possible worlds to fix the semantics of counterfactual conditionals. On Lewis's view, if two possible worlds differ in some facts, or in some laws, they are different worlds. Then, it seems possible to rank worlds according to how *similar* they are. Let's call @ the actual world. In the light of an account of similarity of possible worlds, the truth of a counterfactual conditional $p \,\square\!\rightarrow q$ is defined thus. Given that neither p nor q are true of the actual world, take those possible worlds in which p is true. Call them *p-worlds*. Then, the counterfactual $p \,\square\!\rightarrow q$ is true (in @) if and only if the p-worlds in which q is true are closer to the actual world @ than the *p-worlds* in which q is false.

See **Laws of nature**

Further reading: Goodman (1954); Lewis (1973)

Covering-law model: Expression coined by William Dray (born 1921) to refer to the **deductive-nomological model of explanation** according to which an explanandum is explained by being subsumed under (covered by) a law.

Further reading: Hempel (1965)

Craig's theorem: The logician William Craig (born 1918) constructed a general method according to which given any first-order theory T and given any effectively specified sub-vocabulary O of T, one can construct another theory T' whose theorems are exactly those theorems of T that contain no constants other than those already in the sub-vocabulary O. What came to be known as Craig's theorem is the following: for any scientific theory T, T is replaceable by another (axiomatisable) theory Craig(T), consisting of all and only the theorems of T which are formulated in terms of the observational vocabulary V_0. Craig showed how to construct the axioms of the new theory Craig(T). There will be an infinite set of axioms (no matter how simple the set of axioms of the original theory T is), but there is an effective procedure which specifies them. The new theory Craig(T) is 'functionally equivalent' to T, in that all observational consequences of T also follow from Craig(T). So, for any V_0-sentence O_0, if T implies O_0 then Craig(T) implies O_0. This point was seized upon by instrumentalists, who argued that theoretical commitments in science were dispensable: theoretical terms can be eliminated *en bloc*, without loss in the deductive connections between the observable consequences of the theory.

See **Instrumentalism; Scientific realism; Theoretician's dilemma**

Further reading: Hempel (1965); Psillos (1999)

Critical rationalism: School of thought formed by **Popper** and his followers. It is rationalism since it gives precedent to reason as opposed to learning from experience, but it is critical because it stresses the role of criticism in knowledge. It takes **rationality** to consist in the critical discussion of one's own theory, its subjection to severe tests, its attempted refutation and, should it clash with observations, its rejection. Critical rationalism rejects **inductivism** and bases its own account of **scientific method** on the idea of falsifiability of scientific hypotheses. Unlike traditional **rationalism**, critical rationalism does not take reason (not, of course, experience) to justify beliefs. But it takes it that experience plus deductive logic can falsify beliefs. One of the hurdles it has to jump is the **Duhem–Quine thesis.**

See **Conjectures and refutations; Falsificationism; Musgrave**

Further reading: Miller (1994); Musgrave (1999)

Crucial experiment: It is supposed to distinguish between two competing theories. If theory T_1 entails a prediction P and T_2 entails not-P, a crucial experiment, by favouring either P or not-P, can disprove (falsify) one of the two theories. The idea of crucial experiments was introduced by **Bacon** in his instance of fingerpost. He thought that crucial experiments were essential to his method of **eliminative induction** since they eliminate one of the competing hypotheses about the causes of an effect. Bacon then distinguished between two types of experiments: those that gather data for the development of theories and those that test theories. **Duhem** denied that there any crucial experiments in science. Based on the thought that theories entail predictions only with the aid of auxiliary assumptions, he argued that no experiment can lead to a conclusive refutation of a theory.

See **Duhem–Quine thesis**

Further reading: Duhem (1906); Galison (1987)

Curve-fitting problem: The problem of fitting a curve to the data or, more generally, the problem of fitting a hypothesis to the data. Hypotheses (especially, law-like statements; generalisations) can be represented quantitatively as curves (e.g., as straight lines) plotted in graphs. The data (represented as points on the graph) are always finite. Hence, there is an indefinite number of curves that fit the data. How, then, is a certain curve to be selected as the one that represents the law that binds the data? There are several statistical methods that are used in finding the best-fitting curve (e.g., the method of least squares), but the philosophical problem concerns the cognitive status of the criteria (such as **simplicity**) used to choose the best-fitting curve: what is the connection between these criteria and **truth**? Besides, the philosophical problem is the trade-off that there is bound to be between simplicity and goodness of fit: the more complex the curve is the better it fits with the data, and yet there is always a preference for simpler curves, even if some (perhaps, many) of the data do not lie on the preferred curve.

Further reading: Forster and Sober (1994)

Darwin, Charles Robert (1809–1882): British naturalist, founder of the theory of evolution, according to which natural selection is the driving force of the evolution of species. His celebrated *On The Origin Of Species By Means Of Natural Selection, Or The Preservation Of Favoured Races In The Struggle Of Life*, published on 24 November 1859, was heavily criticised and attacked for promoting atheism by denying the story of the creation of the world, as given in the book of Genesis. His key idea was not so much that evolution occurs but rather that the mechanism by which it occurs is natural selection. According to Darwin's theory, the individuals with the highest **probability** of surviving and reproducing successfully are those that are best adapted to their environment, because of their possession of certain traits. These individuals tend to transmit to their offspring the traits in virtue of which they are adapted to their environment; hence, these traits will increase their frequency in the population and, in the fullness of time, they will prevail. The resulting change in the population is called evolution. What may be called the *Darwinian revolution* was the acceptance of the very idea of evolution (and in particular of Darwin's idea of *branching evolution*, namely, that all organisms descent from common ancestors). The *second* Darwinian revolution took place through the work of Alfred Russel Wallace (1823–1913) and August Weismann (1834–1914), who argued against the inheritance of acquired characteristics. The so-called modern Darwinian synthesis (or neo-Darwinism) is the combination of the two Darwinian revolutions with Mendelian genetics (traced back to the pioneering work of Gregor Johann Mendel, 1822–1884) as the basis of inheritance, and the mathematical theory of population genetics. A central element in the synthesis is the connection between a gene-based understanding of the facts of inheritance with the mechanism of evolution. In the 1950s James D. Watson (born 1928) and Francis Crick (1916–2004) discovered the molecular structure of the gene (the famous double-helix structure of DNA). The genetic understanding of variation strengthened Darwinism because it uncovered the mechanism of natural selection: the production of genetic variation is, by and large, a matter of chance (known as mutation); then there is the process of elimination based on the differential survival and reproduction of organisms and on their adaptedness to their environment. However, natural selection is only one of the **mechanisms** of evolution; other mechanisms are genetic drift and gene flow.

Further reading: Hull and Ruse (1998); Sober (1993); Sterelny and Griffiths (1999)

Darwinism see **Darwin**

Deductive arguments: Logically valid arguments: the premises are inconsistent with the negation of the conclusion. A deductively valid **argument** is such that, if the premises are true, the conclusion has to be true. This property of valid deductive argument (also known as truth-transmission) comes at a price: deductive arguments are not content increasing (the information contained in the conclusion is already present – albeit in an implicit form – in the premises). Though deductive **inference** has been taken to be justified in a straightforward manner, its **justification** depends on the meaning of logical connectives and on the status of logical laws. Those who deny that the laws of logic are **a priori** true argue that deductive inference is justified on broadly empirical grounds. Deductive arguments can be valid without being sound. A sound argument is a deductively valid argument with true premises.

See **Ampliative inference**

Deductive-nomological model of explanation: According to this model, introduced by **Hempel** and Paul Oppenheim, to offer an **explanation** of an event *e* is to construct a valid **deductive argument** of the following form:

Antecedent/Initial Conditions
Statements of Laws
Therefore, *e* (event/fact to be explained)

When the claim is made that event *c* causes event *e* (e.g., that the sugar cube dissolved *because* it was immersed in water), it should be understood as follows: there are relevant laws in virtue of which the occurrence of the antecedent condition c (putting the sugar in water) is nomologically sufficient for the occurrence of the event e (the dissolving of the sugar). A standard criticism of the deductive-nomological model is that, insofar as it aims to offer sufficient and necessary conditions for an argument to count as a bona fide explanation, it fails. There are arguments that satisfy the structure of the deductive-nomological model, and yet fail to be bona fide explanations of a certain event. For instance, one can construct a deductive-nomological 'explanation' of the height of a flagpole having as premises (a statement of) the length of its shadow and (statements of) relevant laws of optics, but this is not an explanation of *why* the flagpole has the height it does. This counterexample relies on the claim that explanations are *asymmetric*: they explain effects in terms of causes and not conversely. There are also bona fide explanations that fail to instantiate the deductive-nomological model. For instance, one can construct an explanation of why there was a car crash (by telling a causal story of how it happened) without referring to any law at all. The joined message of these counterexamples is that the *deductive-nomological* model fails precisely because it ignores the role of **causation** in explanation.

See **Inductive-statistical model of explanation; Laws of nature**

Further reading: Hempel (1965); Psillos (2002); Salmon (1989)

Deductive-nomological-probabilistic model of explanation: Model of **explanation**, introduced by Peter Railton, which tries to reconcile the view that explanations are (at least partly) **deductive arguments** with the view that there can be legitimate explanations of chance events. In explaining a chance explanandum, one constructs a deductive-nomological argument whose conclusion states the **probability** of the explanandum to happen and then one *appends* an addendum which states that the explanandum did obtain. The thrust of the deductive-nomological-probabilistic model is that understanding of why an explanandum occurred does not necessarily consist in producing arguments that show how this event had to be expected with certainty or high probability. The occurrence of the explanandum, be it likely or not, is explained in essentially the same way. If there is a feeling of dissatisfaction with this explanation, it is misguided because it arises from a covert commitment to **determinism**. If we take indeterminism seriously, there is no further fact of the matter as to why a genuinely chancy event happened.

See **Deductive-nomological model of explanation; Mechanism**

Further reading: Railton (1978)

Deductive-statistical model of explanation: It aims to explain statistical regularities, for example, the fact that in a large collection of atoms of the radioactive isotope of Carbon-14 approximately three-quarters of them will very probably decay within 11,460 years. According to **Hempel**, statistical regularities can be explained *deductively* in that their descriptions can be the conclusions of valid **deductive arguments**, whose premises include statistical nomological statements. The deductive-statistical model of explanation is a species of the **deductive-nomological model**, when the latter is applied to the explanation of statistical regularities.

Further reading: Hempel (1965); Salmon (1989)

Deductivism: The view that the only valid arguments are deductively valid arguments, and that deductive logic is the only logic that we have or need. It takes all prima facie non-deductive arguments to be *enthymemes*: arguments with a missing or suppressed premise. After the premise is supplied (or made explicit), the **argument** becomes deductively valid. Consider, for instance, the inductive argument: all observed ravens have been black; therefore (it's probable that) all ravens are black. This is taken by deductivists to be an enthymeme of the following **deductive argument**: all observed ravens are black; nature is uniform; therefore, all ravens are black.

See **Ampliative inference; Musgrave; Popper**
Further reading: Musgrave (1999)

Defeasibility: The feature of ampliative reasoning in virtue of which further information which does not affect the **truth** of the premises can remove the warrant for accepting the original conclusion. Any type of reason that is not conclusive may be called defeasible (in the sense that it is not deductively linked with the conclusion for which it is a reason). To call a warrant (or a reason) defeasible is not to degrade it, *qua* warrant or reason. Rather, it is to stress that (1) it can be defeated by further reasons (or information); and (2) its strength, *qua* reason, is a function of the presence or absence of **defeaters**.

See **Ampliative inference; Justification**
Further reading: Pollock (1986)

Defeaters: Factors (reasons, **evidence** or information) that, when taken into account, can remove the prima facie (defeasible) warrant for a **belief**. On certain analyses of reasoning and warrant (notably the analysis of John Pollock – born 1940), the presence or absence of defeaters is directly linked to the degree to which a subject is warranted to hold a certain belief. Suppose that a subject S has a prima facie (nonconclusive) reason R to believe that Q. Then S is warranted to believe that Q on the basis of R, *unless* either there are further reasons R' such that, were they to be taken into account, they would lead S to doubt the integrity of R as a reason for Q, or there are strong (independent) reasons to hold not-Q. There are two general *types* of defeater that an agent should consider: rebutting defeaters and undercutting defeaters. Suppose that there is some prima facie reason P for believing that Q. A further factor R is called a rebutting defeater for P as a reason for Q if and only if R is a reason for believing not-Q. A factor R is called an undercutting defeater for P as a reason for Q if and only if R is a reason for denying that P offers warrant for Q.

See **Defeasibility; Justification**

Further reading: Pollock (1986)

Definition: The explanation of the meaning of a word by reference to the meanings of other words. The word to be defined is called *definiendum* and the words that define it *definiens*. There has been considerable controversy concerning the existence and philosophical significance of definitions. It has been argued that not all words are definable by reference to other words, since this would lead to an infinite regress. Others (notably **Quine**) have denied the very possibility of definitions since a definition would amount to an analytic **truth** and no truths are analytic. Some think that definitions have no existential implications – to define a concept is not *ipso facto* to be committed to the existence of the *definiendum*. Others argue that there is a kind of definition (what **Mill** called a 'real' as opposed to a 'verbal' definition) that entails the existence of the *definiendum*.

See **Analytic/synthetic distinction; Convention**

Further reading: Quine (1966)

Definition, explicit: It specifies the meaning of a **concept** fully and exhaustively in terms of the meaning of other (already meaningful) concepts. For instance, an explicit **definition** of the term 'father' as 'male parent' offers necessary and sufficient conditions for the application of 'father'. Explicit definability amounts to translatability: an explicitly defined term can be translated into its *definiens*. Hence, it can be replaced in any context in which it occurs by its *definiens*, without loss of content. In the hands of concept empiricists, explicit definition became a device for the specification of the meaning of theoretical terms and dispositional predicates by reference to observational terms and **properties**. They hoped they could show how scientific theories could be true without implying any commitments to **unobservable entities**. **Carnap** was perhaps the first to try to offer explicit definitions of **theoretical terms**. For instance, an explicit definition of the theoretical term 'temperature' would be like this: an object a has temperature of c degrees centigrade if and only if the following condition is satisfied: if a is put in contact with a thermometer, then the thermometer shows c degrees on its scale. This attempt faced a number of problems, both technical and substantial. In any case, not all theoretical terms which scientists considered perfectly meaningful could be explicitly defined. The whole empiricist project presupposed a criterion of meaning based on **verifiability**. This criterion was discredited precisely because it rendered meaningless concepts and statements that scientists have taken to be perfectly meaningful, for example, statements that express **laws of nature**.

See **Concept empiricism; Condition, necessary and sufficient**

Further reading: Carnap (1936)

Definition, implicit: Means for introducing **concepts** and specifying their meaning. They amount to schemata for concepts and their relations. They are associated with **Hilbert**'s axiomatisation of **Euclidean geometry** and his point that the basic concepts that feature in a set of axioms get their meaning from their mutual logical relationships. For instance, concepts like POINT and LINE do not have a meaning independently of the axioms in which they feature. The meaning of the term 'point' is defined by the axioms as whatever is such that any two of those lie on one and only one line. Similarly for 'line' etc. Since the term 'line' appears in the **definition** of the term 'point', and since 'line' has no independent meaning outside the axioms of the system in which it appears, the terms 'line' and 'point', as well as all other terms which appear in the axioms, get their meaning collectively from the axioms in which they feature, that is, from the set of logical relationships they have to all other terms of the axioms.

See **Definition, explicit; Reduction sentences**

Further reading: Horwich (1998a)

Definition, operational: Species of explicit **definition**, introduced by the physicist Bridgman in his defence of **operationalism**. In an operational definition of a concept (e.g., LENGTH, TEMPERATURE) the necessary and sufficient conditions for the *definiendum* specify measurements (or operational procedures). The idea is that the defined **concept** can be meaningfully used in all and only the situations in which the measurement procedures applies.

Further reading: Bridgman (1927)

Degree of belief: Belief may not be an all-or-nothing state; it may admit of degrees. One may believe a proposition to degree r, between zero and one. To measure an agent's degree of **belief** in the **truth** of a proposition (e.g., the proposition that it will rain tomorrow), the agent is made to take a bet on the truth of the proposition. The **betting**

quotient the agent uses measures his/her subjective degree of belief. Hence, degrees of belief are probabilities: they express an agent's subjective **probability** of the truth of a proposition.

See **Bayesianism; Probability, subjective interpretation of**

Further reading: Howson and Urbach (2006)

Demarcation, problem of: The problem of distinguishing between science and non- (or pseudo-)science. **Popper** appealed to falsifiability as the criterion of demarcation. Scientific theories are falsifiable in that they entail testable observational predictions (what Popper has called *potential falsifiers*) which can then be tested in order either to corroborate or to falsify the theories that entail them. Non-scientific claims do not have potential falsifiers. That is, they cannot be refuted. The main obstacle for Popper's criterion comes from the **Duhem–Quine thesis**.

Further reading: Popper (1959)

Descartes, René (1596–1650): French philosopher, physicist and mathematician, to many the founder of modern philosophy. His books include: *Discourse on Method* (1637), *Meditations on First Philosophy* (1641) and *Principles of Philosophy* (1644). His philosophy of science aimed to provide an adequate philosophical foundation to scientific **knowledge**. Feeling the force of the sceptical challenge to the very possibility of knowledge of the world, he tried to show how there could be certain (indubitable) knowledge and, in particular, how science could be based on certain first principles. Knowledge, he thought, must have the **certainty** of mathematics. The vehicles of knowledge were taken to be intuition and demonstration: we can only be certain of whatever we can form clear and distinct ideas or demonstrate truths about. Descartes tried to base his whole foundational conception of knowledge on a single indubitable **truth**, namely, *cogito ergo sum* (I think, therefore I exist). But having demonstrated the existence of God, he took God as a guarantor of the existence of the external world and, ultimately, of its knowledge of it. In *Principia*, he argued that the human mind, by the light of reason alone, could arrive at substantive truths concerning the fundamental **laws of nature**. These, for instance that the total quantity of motion in the world is conserved, were discovered and justified **a priori**, as they were supposed to stem directly from the immutability of God. Accordingly, the basic structure of the world is discovered independently of experience, is metaphysically necessary and known with metaphysical certainty. It is on the basis of these fundamental laws and principles that all natural phenomena are explained, by being, ultimately, grounded in them. How is then empirical science possible? Descartes thought that once the basic nomological structure of the world has been discovered by the lights of reason, science must use hypotheses and experiments to fill in the details. Descartes thought that the less fundamental laws could be known only with moral certainty. His view of nature was mechanical: everything could be explained in terms of matter in motion. Descartes can be seen as favouring the **hypothetico-deductive method**. His reliance on hypothesis was strongly criticised by **Newton**.

See **Causation, transference models; Leibniz; Mechanical philosophy; Rationalism; Scepticism**

Further reading: Descartes (1644); Losee (2001)

Description theories of reference: The reference (or denotation) of a referring expression (e.g., a proper name or a singular term) is specified by means of a description (normally understood as specifying the sense of the referring expression). Since descriptions specify senses, one can understand a description (and hence know the *sense* of a word) without knowing whether this description is true or false (or, more generally, whether it is satisfied or not). This theory of reference has been developed by **Frege** and **Russell**. It was originally assumed that each term is associated with a unique propositional description. But the later Ludwig Wittgenstein (1889–1951) and John Searle (born 1932) advanced the cluster theory: a name refers to whatever entities satisfy a cluster of the descriptions generally associated with it. The description theories came under heavy attack by **Kripke**. He argued that descriptions are neither necessary nor sufficient for fixing the reference of a term.

See **Causal theory of reference; Sense and reference**

Further reading: Devitt and Sterelny (1987); Russell (1912)

Determinism: Intuitively, the view that the past uniquely determines the future. **Laplace** characterised determinism as lawful predictability. He claimed that an intelligence who knew the initial positions and momenta of all bodies in the universe, as well as the laws of motion, could predict their future state of motion with absolute precision. Freed from the epistemic notion of predictability, determinism is taken to be a claim about universal **causation**: each and every event has a fully sufficient nomological condition (i.e., a sufficient cause in accordance with universal laws). Determinism, then, denies the existence of objective **chance** in the world: all events are determined to happen with either **probability** one or zero. Talk about chances is allowed, but only in so far as it expresses our ignorance of the universal laws and/or the initial conditions. The denial of determinism (indeterminism) does not *ipso facto* imply the denial of causal connections among events, since there can be probabilistic (or stochastic) causation. Determinism is supposed to be violated in non-classical physics, but it faces problems in classical physics too, since the Laplacian imagery applies only to closed systems.

See **Probability, classical interpretation of**
Further reading: Earman (1986)

Devitt, Michael (born 1938): Australian philosopher, author of *Realism and Truth* (1984). He has defended common-sense realism and **scientific realism,** taking realism to be a metaphysical position, with two dimensions: *existence* and *independence*. The former states that there are physical things (common-sense objects and entities posited by our scientific theories). The latter secures that the things that exist are independent of the mental and *objective,* where **objectivity** is construed as independence from the knowing subject. Devitt has insisted that no doctrine of **truth** is a constitutive part of the realist position. He has also been a firm defender of **naturalism** and of **nominalism.**

Further reading: Devitt (1997)

Dispositions: Properties of objects in virtue of which they tend to display a characteristic response under suitable circumstances in which the manifestation of the disposition is triggered. Solubility, fragility, elasticity and the like are typical examples of dispositions. For instance, a substance has the disposition to dissolve in water (it is soluble) just in case it manifests this disposition (it dissolves) in the situation in which it is put in water. The very possibility of unmanifested dispositions has rendered dispositions suspect to empiricists. Modern empiricists shifted their attention from dispositional **properties** to dispositional predicates and claimed that the meaning of dispositional predicates is fixed by specifying their correct application to observable situations that involve test conditions and characteristic responses. They favoured a conditional analysis of dispositional predicates. For instance, to say of the predicate 'is soluble' applies to an object is to say that were this object immersed in water it would dissolve. A popular way to specify the meaning of dispositional predicates was via **reduction sentences**. It has been widely supposed that dispositions are sustained by categorical properties. When an object manifests a disposition, when, for instance, a sugar cube put in water manifests its disposition to dissolve in water, its relevant behaviour (its dissolution in water) is caused by some non-dispositionall/categorical property of the object (e.g., the sugar cube's molecular structure). Pressure on this, ultimately reductive, account of dispositions came from the claim that conditional analyses of dispositions fail. It has been argued that even so-called categorical properties entail conditionals; hence this conditional-entailing feature cannot be the mark of a disposition. For instance, it has been argued that the apparently categorical property of something being triangular entails the conditional that, if its corners are counted correctly, the answer must be three. Conversely, it has been argued that dispositions might well fail to entail any conditionals: it might be that the conditions under which a disposition is triggered are such that they cause the loss of the disposition. This is the case of the so-called finkish dispositions. Friends of conditional analyses of dispositions have tried to argue against both kinds of counterexample. But in recent decades, dispositions have been taken seriously as irreducible and causally efficacious parts of the furniture of the world. Dispositions are taken to be intrinsic and occurrent properties whose nature is to tend towards their characteristic manifestation. This view has gained wide currency partly because it resonates with the increasingly popular view that properties are (or are best understood as) **powers**. Pandispositionalism is the view that all properties are purely dispositional in that they need no and admit of no grounding in categorical properties.

See **Essentialism, dispositional; Laws of nature**
Further reading: Mumford (1998)

Duhem, Pierre (1861–1916): French scientist and philosopher of science. In *The Aim and Structure of Physical Theory* (1906), he set the agenda for most subsequent philosophy of science. He tried to offer an account of physics which made it *autonomous*, that is, free of metaphysics. He took metaphysics to be any attempt to offer **explanation** by postulation – that is, explanation in terms of **unobservable** entities and **mechanisms**. Characterisitcally, he took the atomic theory to be such a metaphysical theory. He tried to advance an account of **scientific method** which restricted it to the following: experiments (or observations), mathematics and logic. Duhem took theories to be mathematical tools for the organisation and classifications of phenomena. He thought theories cannot be appraised as true or false, but rather as empirically adequate or inadequate. Hence, he has been taken to be an advocate of **instrumentalism**. Yet, he also perceived that the ability of some theories to yield **novel predictions** cannot be explained by viewing theories instrumentally. He advanced the view that theories aim to offer *natural classifications* of phenomena. He also took as a fundamental postulate of physical theory that it should unify all phenomena under a single system of hypotheses. Duhem is also famous for his view that there cannot be **crucial experiments** in science and that physical theories are tested holistically. He argued that different national characteristics lead to different approaches to science. He favoured French science, which he thought was deep and narrow, over British science, which he took to be broad and shallow. He was particularly hostile to the Maxwellian tradition of building **models** to explain phenomena. He criticised German science for being too geometrical and praised French science for using the analytical style of mathematics.

See **Duhem–Quine thesis; Novel prediction**

Further reading: Duhem (1906)

Duhem–Quine thesis: As **Duhem** first stressed, no theory can generate any observational predictions without the use of auxiliary assumptions. If the prediction is not fulfilled, the only thing we can logically infer is that *either* the auxiliaries *or* the theory is false. Logic alone cannot distribute blame to the premises of an **argument** with a false conclusion. This means that one can always attribute falsity to the auxiliaries and hold on to the theory come what may. **Quine** extended Duhem's thesis by stressing that our overal theory of the world (which includes logic as well as mathematics and geometry) faces the tribunal of experience as a whole. Should a conflict with experience arise, one can modify or abandon *any* part of theory (including logic and mathematics) in order to accommodate the recalcitrant experience. Accordingly, there is no statement which is immune to refutation. Quine used this thought to argue that there are no analytic, or synthetic a priori, statements. The revisions to our overal theory of the world are governed by several pragmatic principles, such as the **principle of minimal mutilation** and considerations of **simplicity**. The Duhem–Quine thesis has been suggested as an algorithm for generating empirically equivalent theories: for any evidence and any two rival theories T and T', there are suitable auxiliaries A such that T' & A will be empirically equivalent to T (together with its own auxiliaries). Hence, it is argued, no evidence can tell two theories apart. There is no proof that *non-trivial* auxiliary assumptions can always be found. But suppose it were true; what would this show? Since the Duhem–Quine thesis implies that any theory can be saved from refutation, it does create some genuine problems to **falsificationism**. But it does not create a similar problem to **inductivism**. From the fact that any theory can be suitably adjusted so that it resists refutation it does not follow that all theories are equally well confirmed by the evidence.

See **Holism, confirmational; Underdetermination of theories by evidence**

Further reading: Duhem (1906); Quine (1975)

Dummett, Michael (born 1925): British philosopher, one of the most influential thinkers of the twentieth century, author of *Truth and Other Enigmas* (1978) and *The Logical Basis of Metaphysics* (1991). He worked on **Frege's** philosophy of language and mathematics and devised a modern form of anti-realism based on the idea that **truth** is not evidence transcendent. He resuscitated **verificationism** and defended intuitionism in mathematics – the view that equates truth and proof and denies the principle of bivalence, namely, that every well-defined statement is either true or false.

See **Putnam; realism and anti-realism; scientific realism**

Further reading: Dummett (1991); Weiss (2002)

Dutch-book see **Bayesianism; Conditionalisation**

Dutch-book theorem see **Bayesianism**

Earman, John (born 1942): American philosopher of physics, author of *A Primer on Determinism* (1986) and *Bayes or Bust* (1992). He has contributed to the philosophy of **space** and **time**, the philosophy of **quantum mechanics**, and broader methodological issues such as the theory of **confirmation**. He has tried to move between the horns of traditional philosophical dichotomies (e.g., the substantivalism/relationalism debate about spacetime or the Bayesian and non-Bayesian theories of confirmation) stressing that adequate philosophical theories can be developed only by utilising resources and insights from both sides of the traditional dichotomies.

Further reading: Earman (1986, 1992)

Einstein, Albert (1879–1955): German-born American physicist, arguably the most important scientist of all time, founder of the special and the general theories of relativity. In 1905, his *annus mirabilis*, he saw published his paper on Brownian motion, in which he made a case for the existence of atoms; his paper on the photoelectric effect, in which he posited the existence of photons; and his paper 'On the Electrodynamics of Moving Bodies', in which he laid the foundations of the Special Theory of Relativity. According to the principles of this theory, (1) the laws of physics are the same in all inertial frames of reference and (2) the speed of light is the same for all observers, no matter what their relative motion. This theory rescued Maxwell's equations by making deep changes to the fundamental conceptual framework of Newtonian mechanics. **Space** and **time** were relativised and (in the work of Minkowski) were united to a four-dimensional manifold: **spacetime**. His account of simultaneity, which was presented in terms of synchronisation of clocks and observers, led several of his followers to think that he espoused **operationalism** and **positivism**. But this is exaggerated. Einstein was a realist about the structure of spacetime and, in particular, defended the invariance (and reality) of the spacetime interval. In 1915, he published his work on the General Theory of Relativity, which extends the insights of the special theory to gravity. The notion of frame of reference was extended so that it included accelerated frames. According to the Principle of Equivalence, a frame of reference that falls freely in a uniform gravitational field is equivalent to an inertial frame. This principle, whose satisfaction implied that space could not be Euclidean, led Einstein to develop a theory of gravity that involves curved spacetime. According to his field equation, the curvature of spacetime depends on the stress-energy within the spacetime. One important consequence of the General Theory is that light would be affected by a gravitational field much more than Newton had predicted (and hence that light rays would bend near massive bodies). This prediction was confirmed by Arthur Eddington (1882–1944) in 1919, a fact that led to the wider acceptance of Einstein's theory. Thinking about theories, Einstein drew a distinction between constructive theories (which rest on **models** of the phenomena) and principle theories, which start with general principles. He thought that his own theories of relativity belonged to the latter category.

See **Bohr; Quantum mechanics, interpretations of; Thought experiment**

Further reading: Fine (1986); Zahar (1989)

Eliminative induction: Mode of **induction** based on the elimination of rival hypotheses. It was championed by **Bacon** and is akin to **Mill's methods**. It is not so much concerned with how hypotheses are generated as with how they are justified once they become available. By eliminating all but one available hypothesis that stands in a certain relationship with the **evidence** (e.g., they entail the evidence, or they explain it etc.), the one that remains is taken to be likely to be true. This judgement depends on the further assumption that the **truth** is, or is likely, to be among the already available hypotheses. This may or may not be a warranted assumption, depending on the context. It has been argued that it is mostly explanatory considerations that govern the elimination of hypotheses (hypotheses that are eliminated offer no or poor **explanation** of the evidence). Then, eliminative induction is a species of **inference to the best explanation**.

See **Enumerative induction; Mill**

Further reading: Lipton (2004); Salmon (1967)

Ellis, Brian (born 1929): Australian philosopher of science, author of *Truth and Objectivity* (1990) and *Scientific Essentialism* (2001). He has defended an epistemic conception of **truth** and has tried to show how this is compatible with **scientific realism**. More recently, he has defended **dispositional essentialism**.

Further reading: Ellis (2001)

Emergence: The process by which novel **properties** of systems (or complexes) arise. It is supposed to characterise the relationship between a whole and its parts: the whole has novel properties vis-à-vis its parts and the laws that govern their interactions. This idea of novel properties is explained in various ways. For instance, it is said that they are non-deducible or non-predictable from the properties of the parts of a system, or that they are non-reducible to the properties of the parts of the system. But their distinctive feature is that they are supposed to have novel causal **powers**. Emergentism is the view that nature has a hierarchical (multi-layered) organisation such that the higher layers (though probably composed of elements of the lower layers) are causally independent of, and irreducible to, the lower layers. Though currently popular in the philosophy of mind, emergentism has a long history that goes back to the relations between biology and chemistry, on the one hand, and physics, on the other, in the beginning of the twentieth century. The British Emergentists were a group of scientists and philosophers – including C. D. Broad (1887–1971), C. Lloyd Morgan (1852–1936) and Samuel Alexander (1859–1938) – who argued that the task of the sciences other than fundamental physics was to account for the emergent properties and their causal and nomological behaviour. They were committed to the existence of downward **causation,** namely, the view that the novel powers of the emergents causally influence the behaviour of entities at the lower levels.

See **Reduction; Vitalism**

Further reading: MacLaughlin (1992)

Empirical adequacy: Property of theories in virtue of which they save the phenomena. A theory is empirically adequate if and only if all of its observational consequences are true. Being an advocate of the **semantic view of theories, van Fraassen** cast the requirement of empirical adequacy in model-theoretic terms. For a theory to be empirically adequate it should be the case that the **structure** of appearances is embedded in one of the models of the theory (i.e., that the structure of appearances is isomorphic to an empirical sub-structure of a model of a theory). This way of casting the requirement of empirical adequacy frees it from the commitment to a distinction between observational and theoretical vocabulary. Commitment to empirical adequacy is suitable for **instrumentalism** since a theory may be empirically adequate and false. According to **constructive empiricism,** empirical adequacy replaces **truth** as the aim of science.

Further reading: van Fraassen (1980)

Empirical equivalence see **Underdetermination of theories by evidence**

Empiricism: The view that experience is the only source of information about the world. Though many empiricists have taken this claim to be constitutive of empiricism, this way of putting the view makes it a factual claim about the genesis of **knowledge**, and it may be best to characterise empiricism as the view that experience is (ought to be) the only source of **justification** for substantive claims about the world. Empiricism is the rival of **rationalism**. Interestingly, *empirics* were called a post-Hippocratic school of medicine, under the leadership of Philinos of Cos and Serapion of Alexandria, which claimed that all medical knowledge arises out of: (1) one's own observations; (2) the observations of others; and (3) analogical reasoning. They were opposed to *dogmatists*. **Bacon** compared empirics with ants (since they collect only experimental results) and dogmatists with spiders ('who make cobwebs out of their own substance'). His own alternative (his *new* empiricism) was compared to bees: the experimental data were transformed to knowledge by reason, following the **scientific method**. Empiricism took its modern form with **Locke, Berkeley** and **Hume**. Yet, their disagreement on a number of issues (are there abstract ideas? can we distinguish between primary and secondary qualities? can things exist unperceived? can there be causal knowledge?) highlights the fact that empiricism is far from being a solid and tight doctrine. However, we can say that empiricism is characterised by the rejection of synthetic **a priori** knowledge and by a disdain towards metaphysics – since the latter is supposed to transcend experience and whatever can be known on its basis. **Leibniz** famously claimed that we are all empiricists in 'three-quarters of our actions', but he took the fourth quarter (namely, the knowledge of first principles and in particular the knowledge of necessary truths) to require the adoption of other (non-empirical) modes of knowing. The empiricist camp has been divided over this matter. Though there is unanimity that there can be no substantive knowledge of the world by the lights of reason only, some empiricists (notably **Mill** and **Quine**) have taken the view that all truths (even the truths of logic and mathematics) are synthetic and a posteriori, while others (notably **Carnap** and other followers of **logical positivism**) have taken the view that there is a special category of non-empirical truths which are knowable a priori – but they are analytic truths and hence do not require a special faculty of rational insight or intuition. Among the radical empiricists who take all knowledge to be a posteriori, there are those (like Mill) who think that all knowledge arises out of experience by means of **induction** (and it is justified on this basis) and those (like Quine) who take experience to regulate a system of beliefs by exerting negative control on it – when there is conflict between the system of beliefs and experience, there must be suitable adjustments to this system to restore coherence, governed by general principles such as the **principle of minimal mutilation**. Empiricists have disagreed over: the exact limits of experience (do they include whatever is actually observed or whatever is observable, and, if the latter, observable by whom? Me, us, any human being, God?); the legitimacy and the scope of the methods which start from experience (is induction justified? If not, is **scepticism** inevitable for empiricists? Is reasoning by **analogy** legitimate and can the analogy be extended to entities that cannot be experienced, e.g., to unobservable entities?); the content of experience (is this composed of **sense data** or are material objects directly experienced?) It might then be best to talk of empiricisms, united by a call to place experience firmly at the heart of our cognitive give-and-take with the world.

See **Concept empiricism; Constructive empiricism; Given, the; Judgement empiricism; Logical positivism; Neo-Kantianism; Reductive empiricism; Scientific realism**

Further reading: Ayer (1936); Carnap (1936); Mill (1911); Quine (1951); Reichenbach (1938); Solomon (2001); van Fraassen (1985)

Entity realism: Form of **scientific realism** according to which one may accept the existence of all sorts of theoretical entities (e.g., electrons, genes, Higg particles etc.), while one may reject the high-level theories in which descriptions of these entities are embedded. It has been entertained by **Hacking** and **Cartwright**. A major motivation for entity realism comes from laboratory life: experimenters have good reasons to believe in specific **unobservable entities,** not because they accept the relevant theories, but rather because they *do* things with these entities. These phenomena of laboratory life would be inexplicable if these entities did not exist. As Hacking has famously put it regarding quarks: 'So far as I'm concerned, if you can spray them, then they are real.' Cartwright bases her entity realism on an inference to the likeliest cause.

See **Structural realism**

Further reading: Cartwright (1983); Hacking (1983)

Enumerative induction: Mode of **induction** based on the following: if one has observed *n* As being B and *no* As being not-B, and if the evidence is enough and variable, one should infer that (with high probability) 'All As are B'. For obvious reasons, it can be called the more-of-the-same method. The basic substantive assumptions involved in this ampliative method are that: (1) there are projectable regularities among the data; and (2) the pattern detected among the data (or the observations) in the sample is representative of the pattern (regularity) in the whole relevant population.

See **Ampliative inference; Eliminative induction; Straight rule of induction**

Further reading: Salmon (1967)

Error-theory: Anti-realist view of certain domains of discourse (sets of propositions, theories etc.) according to which a certain assertoric discourse is in massive error, since there are no entities of the type required for this discourse to be true. Take a certain set D of propositions (e.g., ethical or mathematical or modal). Error-theorists claim that the propositions in D should be taken at face value, that is, as implying the existence of certain kinds of entities, but add that the propositions in D are *false*: there are no entities or facts that make them true. Examples of error-theoretic anti-realism are Field's mathematical fictionalism and Mackie's subjectivism in ethics.

See **Fictionalism, mathematical; Literal interpretation; Quasi-realism; Realism and anti-realism; Truth**

Further reading: Wright (1992)

Essentialism: Metaphysical view, going back to **Aristotle** and resurfacing after **Kripke**, according to which there is a sharp distinction between essential and accidental **properties**: an object is what it is in virtue of its essential properties, which it holds necessarily. This view had become very unpopular until fairly recently, partly because of the criticism, stressed by Wittgenstein and the logical positivists among others, that the only necessity is verbal necessity. **Quine**'s critique of the **analytic/synthetic distinction** was taken to discredit essentialism even further: if there is no sharp distinction between analytic truths and synthetic truths, essentialism cannot even be considered as a claim about a special subset of truths – the analytic ones. Quine did not just deny *de dicto* necessity (understood as analyticity). He also denied *de re* necessity. The latter is supposed to be necessity *in* the world. Quine argued that this kind of distinction between essential and accidental properties cannot be drawn. A mathematician is essentially rational and accidentally bipedal. A bicyclist is essentially bipedal and accidentally rational. What, then, are the essential and the accidental properties of a *bicyclist mathematician*? However, developments in modal logic and the possible world semantics, and Kripke's disentangling of the modal status of a proposition from its epistemic status, have made essentialism credible again.

See **Essentialism, dispositional; Hull**
Further reading: Bealer (1987)

Essentialism, dispositional: The view that **natural kinds** (or natural properties) have dispositional essences, that is, causal **powers** that they possess *essentially* and in virtue of which they are disposed to behave in certain ways. For instance, water has essentially the power to dissolve salt and it is in virtue of *this* power that it does dissolve salt in the actual world and that it is a *necessary truth* that water dissolves salt. **Laws of nature** are ontologically dependent on the intrinsic natures (essences) of natural kinds: given that the natural kinds are essentially disposed to behave in certain ways, the causal laws they give rise to are fixed. This view challenges the Humean assumption that laws of nature supervene on non-modal facts.

See **Dispositions; Essentialism**
Further reading: Ellis (2001)

Ethics of science: Emerging discipline in the border between ethics and philosophy of science. Like ethics in general, it is divided into three areas: meta-ethics; normative ethics and applied ethics. The meta-ethical enterprise is concerned with the status of ethical norms that may operate in science: where do they get their justification from and what is their status? The normative ethics of science is concerned with the development of ethical theories about the proper conduct of scientific research. For instance, it has been suggested that there is a core of ethical principles (or rules) that ought to constrain scientific research – for example, principles about misconduct (plagiarism, falsification of data etc.). Applied ethics of science is concerned with practical ethical problems (sometimes called ethical dilemmas) that may occur in science. Some of the principles that have been suggested as part of a moral theory of scientific research are: the principle of public responsibility (which encourages researchers to inform the public about the consequences of their research); the principle of honesty (which condemns fraud); the principle of credit (which encourages that credit should be given to all those who have contributed to the research); the principle of respect of human subjects and others. Important issues arise when attention is focused on the universality of these principles, the resolution of conflict that might arise when they are applied and their relations to the epistemic goals of science.

Further reading: Resnik (1998)

Etiological explanation see **Functional explanation**

Euclidean geometry: Geometrical system in which Euclid's fifth postulate holds. It was presented in an axiomatic form for the first time by Euclid (365–300 BCE), a Greek geometer author of *Elements*. He started with five postulates (axioms), the fifth of which (known as the parallel postulate) states that from a point outside a line exactly one line parallel to this can be drawn. The first complete axiomatisation of Euclidean geometry was by **Hilbert** in 1899. Euclidean geometry is supposed to be the geometry of the physical (flat) **space** as we experience it and **Kant** thought it was constitutive of the form of spatial intuition. In the nineteenth century, **non-euclidean** geometrical systems emerged which denied the parallel postulate.

Further reading: Torretti (1978)

Events: According to the standard view, advocated by Davidson, events are spatio-temporal **particulars,** which can be described in different ways. A certain event, for instance, can be properly referred to as the breaking of the vase on the table. But it can also be referred to as the breaking of the Jones's wedding present. Events *qua* happenings in the world should not be confused with their descriptions. Their descriptions can be partial, perspectival or incomplete and the same event can be referred to by means of different descriptions. The description of an event is a means to identify it, but it is the event itself and not its description(s) that enter into causal relations. An important alternative approach, due mostly to Jaegwon Kim (born 1934), is that events are exemplifications of **properties** by objects at particular times. So, an event is a triple [x, P, t], which states that the property P is exemplified by the object x at time t. An advantage of this account over Davidson's is that it makes clear how *properties* can be causally efficacious. Generally, we can talk about event-tokens (i.e., particular events, like the smashing of the pink vase by John at 12 noon on Thursday, 13 October 2005 in Zurich) and event-types (i.e., events generically understood, e.g., a smashing of a vase).

See **Causal relata; Causation**

Further reading: Davidson (1980)

Evidence: Narrowly understood, any kind of observation, observational report, experiential input, empirical information, or datum that can be used to support or discredit a hypothesis or theory. Broadly understood, whatever information or reason can be adduced in favour of or against the **justification** of a **belief.** In philosophy of science, typically, the concept of evidence is understood narrowly. Hence, all evidence is taken to be empirical or observational. Some philosophers distinguish between three concepts of evidence: classificatory, comparative and quantitative. In the first case, the issue is whether some observation is evidence (i.e., confirms or supports) a theory or hypothesis. In the second case, the issue is whether some observation is evidence for a certain hypothesis more than it is for another (i.e., if it confirms one hypothesis more than it confirms another). In the third case, the issue is the *degree* to which an observation is evidence for a hypothesis (i.e., the degree to which it confirms a hypothesis). The relationship between evidence and theory is the object of the theory of **confirmation**. When the evidence is sufficient for the **truth** of a hypothesis, the evidence is conclusive. When the evidence is not sufficient to establish the truth of a hypothesis, it is inconclusive or prima facie. Inconclusive evidence can nonetheless be strong enough to warrant belief. An important trend in the philosophy of science (with as disparate advocates as Popperians and Bayesians) takes it that the real issue is not whether evidence supports, or warrants belief in, a hypothesis, but rather how beliefs are adjusted (changed, abandoned, modified) in the light of new evidence – that is, in the light of new information we come to accept as true.

See **Bayesianism; Old evidence, problem of**
Further reading: Achinstein (2005)

Evolution see **Darwin**

Evolutionary epistemology: Approach to epistemology which aims to apply evolutionary **mechanisms,** such as blind variation and selective retention, or genotype/phenotype pairs, to epistemological issues, and in particular to how **beliefs** (or **concepts,** or theories) are formed, evaluated, changed or overthrown. Though this research programme has been quite fruitful, traditional epistemologists have dismissed it as being purely descriptive and hence irrelevant to epistemology. The claim that the mechanisms or organs humans use to interact with the world, and hence to form beliefs, have been shaped by biological evolution is generally accepted. This view has been called *Evolutionary Epistemology of Mechanisms*. The stronger claim, namely, that theories, concepts and beliefs are subjected to such an evolution is much more contentious, though it has been defended by many, including **Popper** and the American social scientist Donald Campbell (1916–1996). This view has been called *Evolutionary Epistemology of Theories – EET*. Some advocates of *EET* take the biological model of the growth of scientific **knowledge** merely as an **analogy**, while others take it quite literally. Among the former, **Hull** has developed a selectionist account of scientific concepts and theories based on conceptual lineages emulating biological lineages.

See **Darwin; Naturalism**

Further reading: Campbell (1974); Hull (1988)

Experiment see **Crucial experiment; Thought experiment**

Explanation: An answer to a why question. The explanation of a fact (explanandum) is achieved by stating some causal-nomological connections between it and the facts that are called upon to do the explaining (explanans). There are two broad views as to the nature of explanation. First, explanations are **arguments**: to explain an event is to construct an argument such that (a description of) the explanandum follows (logically or with high **probability**) from certain premises which state **laws of nature** (either universal or statistical-probabilistic) and initial conditions. Most typical species of this genus are the **deductive-nomological** and the **inductive-statistical** models of explanation. Second, explanations are *not* arguments: they are causal stories as to how the explanandum was brought about. On this view, an explanation need not cite any laws to be complete; it is enough that it specifies the causal **mechanisms** at work or that it gives some portion of the causal history of the explanandum. This view has also been called the ontic conception of explanation and has been advocated by **Salmon**. It takes explanation to be intimately linked to **causation**. Explanation is then seen as the process in virtue of which the explananda are placed in their right position within the causal structure of the world. A view consistent with both approaches is that explanation has to do with understanding and that understanding occurs when we fit the explanandum into the causal-nomological nexus of things we accept.

See **Functional explanation**

Further reading: Hempel (1965); Psillos (2002); Salmon (1989)

Explanation, causal: Explanation of why something happened by citing its causes. Two important questions concerning causal **explanation** are: Does all explanation have to be causal? Does all causal explanation have to be nomological? There are philosophers who think that there are non-causal explanations (e.g., mathematical explanation, or explanation by reference to conservation laws, or to general non-causal principles). More interestingly, the explanation of less fundamental laws by reference to more fundamental ones (and the very idea that explanation amounts to **unification**) is said to be non-causal. Given, however, that there are genuine cases of causal explanation, the further issue is whether there can be singular causal explanation, that is, causal explanation that does not make reference to **laws of nature**, be they universal or statistical.

See **Causation; Causation, singular**

Further reading: Lewis (1986); Psillos (2002)

Explanation, mechanistic see **Causal process; Mechanism**

Explanation, pragmatics of: Those aspects of **explanation** that relate to the *act* or the *process* of explaining, instead of the *product* of explanation. An explanation is seen as an answer to a why question and the *relevant* answers are taken to depend on the presuppositions or the interests of the questioner, on the space of alternatives and, in general, on the context of the why question. Here is one famous example. A priest asks Willie Sutton, when he was in prison, '*Why did you rob banks?*', to which Sutton replied: '*Well, that's where the money is.*' The thought here is that this is a perfectly legitimate answer for Sutton, because for him the space of relevant alternatives (the contrast class) concerns robbing groceries or diners or petrol stations etc. But the space of relevant alternatives for the priest is quite different: not robbing anything, being honest etc. The difference of perspective can be brought out by placing the emphasis on different parts of the question, 'why did you rob *banks*?', as opposed to 'why did you *rob* banks?'. Pragmatic theories of explanation, very different in their details but quite similar in their overal focus on the act of explaining and the contrast classes, have been offered by **Achinstein**, Alan Garfinkel (born 1945) and **van Fraassen**.

Further reading: Garfinkel (1981); van Fraassen (1980)

Explanation, teleological see **Functional explanation**

Explanation, unification model of: According to a long-standing philosophical tradition, the **explanation** of a set of laws amounts to their being unified into a comprehensive nomological system. Explanatory unification is achieved by showing how descriptions of empirical laws are derived within a theoretical system, whose axioms capture the fundamental **laws of nature**. Though explanation is taken to be deductive derivation, it is derivation within a maximally unified theoretical system, that is, a theoretical system that employs the smallest possible number of axioms (fundamental laws) to account for the largest possible number of less fundamental laws. If a large number of seemingly independent regularities is shown to be subsumable under a few comprehensive laws, our understanding of nature is improved since the number of laws that have to be taken as unexplained explainers is minimised. Yet, the concept of **unification** resists a fully adequate **explication**.

See **Unity of Science**

Further reading: Kitcher (1989)

Explication: Analytical procedure, suggested by **Carnap**, by means of which an ordinary imprecise **concept** is made more precise. The *explicandum* is the concept to be explicated while the *explicatum* is the concept (or concepts) that sharpens the content of the *explicandum*. For instance, the explication of the concept WHALE involves, as its *explicatum*, the concept MAMMAL since, though whales are aquatic animals, they are mammals and not fish. The explication of a concept does not necessarily lead to a single *explicatum*. When, for instance, Carnap applied this procedure of explication to the pre-scientific concept of **probability**, he suggested two *explicata*. The first, probability$_1$, is the concept of logical probability that takes probabilities to be (rational) **degrees of belief** in propositions. The second, probability$_2$, is the concept of objective probability that identifies probability with the relative frequency of an event in a certain sequence of events.

Further reading: Carnap (1950b)

External/Internal questions: Distinction introduced by **Carnap**. He suggested that questions concerning the existence of certain kind of entities (e.g., 'are there numbers?', 'are there **properties**?', 'are there classes?' etc.) can be understood in two different ways: either as external questions or as internal ones. *External* questions are meant to be metaphysical: they concern the existence or reality of the system of entities as a whole. Answering such questions presupposes that the existence of the relevant entities can be asserted or denied independently of a certain language. Carnap took this thought to be fundamentally wrong. No metaphysical insight into their nature is needed for the introduction of a new kind of entities. All that is needed is the adoption/construction of a certain linguistic framework whose linguistic resources make it possible to talk about such entities. Once the framework is adopted, questions about the existence or **reality** of the relevant entities lose any apparent metaphysical significance. They become *internal*: that certain entities exist follows from the very adoption of the framework. No facts in the world will force us to adopt a certain framework. The only relevant considerations are pragmatic: the efficiency, fruitfulness and **simplicity** of each proposed linguistic framework.

See **Analytic/synthetic distinction**
Further reading: Carnap (1950a)

Fallacy: An erroneous inferential pattern. Formal (deductive) fallacies are patterns that appear to have the form of a **deductive argument** but are logically invalid. For instance, the fallacy of affirming the consequent has the invalid logical form {if p then q; q; therefore p}. Informal fallacies are reasoning patterns that appear to provide good or strong reasons for a certain conclusion but fail to do so.

The fallacy of equivocation, for instance, is the result of using an ambiguous word with different senses within the same argument.

See **Post hoc, ergo propter hoc**

Further reading: Engel (2000)

Falsificationism: The view, advocated by **Popper**, that we needn't despair if **inductivism** fails! **Scientific theories** can still be falsified by the **evidence**. It rests on the asymmetry between verification and falsification. The tests of scientific theories are attempts to refute a theory. Theories that survive severe testing are said to be corroborated. But, according to falsificationism, no amount of evidence can inductively support a theory. Advocates of falsificationism have not managed to come to terms with the **Duhem–Quine thesis**.

See **Corroboration**

Further reading: Popper (1959)

Feigl, Herbert (1902–1988): Austrian-American philosopher, member of the **Vienna Circle**. He founded the Minnesota Center for Philosophy of Science in 1953. He was one of the architects of the liberalisation of **logical positivism**. He criticised **verificationism** and claimed that it confuses the issue of what constitutes evidence for the **truth** of an assertion with the issue of what would make this assertion true. He defended the compatibility of **empiricism** with **scientific realism**. For him, something is real if it is required in the coherent spatio-temporal-causal account of the world which science offers. This, he thought, gave him a solid conception of empirical realism in contradistinction to metaphysical realism. He defended the **rationality** of **induction** against sceptical attacks to it and drew an important distinction between the validation and the vindication of an inferential method. Feigl became famous for his defence of the identity theory of mind (the view that mental properties are physical – neurophysiological – properties). He took the identity of mental and physical properties to be an a posteriori theoretical identity, justified on the basis of how well it explains the facts.

See **Induction, the problem of; Validation vs vindication**

Further reading: Feigl (1981)

Feminist empiricism: Feminist approach to science that starts with a criticism of the traditional model of science as a fully objective and value-free enterprise. It has been part of the naturalist turn in the philosophy of science with an emphasis on the role of social factors on science. It started as the 'spontaneous consciousness' of feminist scientists (especially in biology and the social sciences) who criticised gender bias in science as producing 'bad' science. Hence, it was more of a call at reform (aiming at the improvement) of existing practices in science than a call for a radical critique and change of them. Though it does not deny that logic and nature impose constraints on our theorising about the world, it claims that **knowledge** is always situated, local, perspectival and social.

See **Empiricism; naturalism**

Further reading: Longino (1990)

Feminist philosophy of science: Philosophical engagement with science from a perspective that focuses on gender issues and their role in, and implications for, science. It calls into question any attempt to address traditional philosophical questions and problems (e.g., the **objectivity** of scientific **knowledge**, the **justification** of **scientific method** etc.) that presents itself as being universal and objective without taking firmly into account the interests and the point of view of women. More recent trends focus on broader social and political issues and stress the importance and indispensability of taking into account particular contexts (gender being one of them) in thinking about science and its claims to **objectivity**. A key thought in feminist approaches to science is that feminist philosophy of science should be seen as an *active* attempt to liberate our conceptual categories from gender biases, to criticise and remove relations of power and domination in science and life and to expand democracy in the production and use of knowledge.

See **Feminist empiricism; Feminist standpoint**
Further reading: Alcoff and Potter (1993)

Feminist standpoint: Feminist approach to science with affinities to a Marxist perspective on epistemological issues. It focuses on how gender differences shape or constrain what can be known and how. The feminist standpoint gives priority to the lives, experiences and values of women. Women, being outside the network of power and the dominant institutions, are said to have a more objective understanding of what goes on in society and less interest in preserving ignorance. They are said to have a clearer picture of the social reality and of what needs to be done to change. This makes the feminist standpoint the locus of **objectivity**. **Harding** has called this 'strong objectivity'. It separates the demand of objectivity from the demand for neutrality (or disinterestedness) and claims that situated **knowledge** (and in particular knowledge that starts from the lives and needs of marginalised subjects) can be objective.

Further reading: Harding (1986)

Feyerabend, Paul (1924–1994): Austrian-American philosopher, author of *Against Method* (1975). He started his career as an advocate of **critical rationalism,** but he became famous for his later epistemological anarchism, the view encapsulated in the claim that there is no such thing as *the* scientific method. His often misunderstood slogan 'anything goes' is not meant to suggest a methodological principle that one should abide by; rather, as he thought, this kind of statement is the only useful generalisation about the scientist's conception of the **scientific method** that can be drawn from the history of science. Feyerabend became known for his view that all observation is theory laden. He also advocated meaning holism and defended some version of **incommensurability**.

Further reading: Feyerabend (1975); Preston (1997)

Fictionalism: The view regarding a set of putative entities that they do not exist but that they are (useful) fictions. On this view, to say that one accepts the proposition that *p as if* it were true is to say that *p* is false but that it is useful to accept whatever *p* asserts as a fiction. This stance was introduced by **Vaihinger**.

Further reading: Field (1980); Vaihinger (1911)

Fictionalism, mathematical: The view that there are no numbers (or other mathematical entities), but that mathematics is still useful, since numbers and other mathematical 'entities' are useful fictions. Fictionalism is a kind of **instrumentalism** about mathematics. In the last few decades, it has been defended by Hartry Field (born 1946). One important motivation for fictionalism is traditional **nominalism** and its aversion to **abstract entities**. Being in the nominalist camp, mathematical fictionalism opposes mathematical **Platonism**. Fictionalism points to the existence of important ontological and epistemological problems with abstract entities. To the standard Platonist argument that commitment to mathematical entities is indispensable for doing science, fictionalists counter

that mathematics is *dispensable*. To show this, fictionalists embark on a nominalisation programme: they try to
show that any physical theory T which uses mathematical
vocabulary can be replaced with another physical theory
T' which has exactly the same nominalistic (i.e., abstract-
entities-free) consequences as T, but is mathematics
free. Hence, if fictionalists about mathematics assert that a
physical theory is true, they mean that its nominalised version is true. If mathematics is false, how can it be useful
to science? Fictionalists claim that mathematics is useful
because, being a conservative extension of mathematics-
free (that is, nominalistic) scientific theories, it facilitates
deductions that could be, in principle, performed within
a nominalistic theory. Hence, the fictionalist credo is that
mathematics is a body of *useful* fiction. Fictionalism has
been challenged on many grounds. The most important
ones are based on arguments that aim to show that mathematics is not really conservative; on arguments that question the generalisability of the nominalisation programme
(e.g., can quantum mechanics be nominalised?); and finally on arguments that question the distinction between
the mathematical vocabulary and the physical vocabulary.

Further reading: Field (1980); Shapiro (1997)

Fine, Arthur (born 1937): American philosopher of physics
who has worked on **Einstein**'s philosophy of science. He
has authored *The Shaky Game: Einstein, Realism and
the Quantum Theory* (1986). He has had a considerable
impact on the realism debate by advancing and defending
the **natural ontological attitude**. More recently, he has
worked on **pragmatism** and **fictionalism**.

Further reading: Fine (1986)

Fodor, Jerry (born 1935): Very influential American philosopher of mind, with important work on issues pertaining

to the philosophy of science, especially **reduction**, the **theory-ladenness of observation**, and **holism**. He is the author of *Psychological Explanation* (1968) and *Concepts* (1998). Fodor has argued against holism and has defended the view that observation has a kind of independence from theory. By advancing a modular theory of the mind, he argued that the perceptual module is informationally encapsulated, that is, it can process information in a way that is not influenced or examined by other processes. What follows is that perception is not theory-laden in the way it has been standardly assumed. Theories, by being inaccessible to perceptual modules, do not affect the way perceivers see things. Hence, even if scientists may work with different theories, they may well see the world in exactly the same way.

See **Reduction**

Further reading: Fodor (1974, 1998)

Formal mode vs material mode: Distinction introduced by **Carnap** to mark the difference between expressions that are meant to refer to the language (better, the syntax) and expressions that are meant to refer to the world. Whereas the statement 'A table is a thing' is in the material mode, the statement ' "Table" is a thing-word' is in the formal mode: it refers to the linguistic properties of a word. This distinction lay at the heart of Carnap's logic of science; it captured his thought that philosophy of science should be concerned with the logical analysis of the language of science.

Further reading: Carnap (1928)

Foundationalism: Hierarchical-linear theory of **justification**. **Beliefs** are divided into two categories: basic (which are self-justified or self-evident) and derived (which depend on the basic beliefs and whose justification is inferential). Foundationalist approaches are either rationalistic or empiricist. Empiricists typically take the content of basic beliefs to be phenomenal (about **sense data**, whose presence seems indubitable). Rationalists focus their attention on innate ideas and beliefs that are produced by introspection – which are supposed to be indubitable. However, it has been claimed that there are no beliefs that are basic (and hence, indubitable). Even if it were granted that *some* beliefs are basic, the further problem remains of the legitimacy and justification of the methods that are supposed to transfer justification from the basic to the derived beliefs. The logical positivists were supposed to be advocates of foundationalism, though their debate over **protocol sentences** shows that they have had a rich and nuanced conception of the alleged foundations of **knowledge**.

See **Coherentism; Given, the**

Further reading: Chisholm (1982); Williams (2001)

Frege, Gottlob (1848–1925): German mathematician and philosopher, the founder of modern mathematical logic and one of the most influential figures in analytic philosophy. In *The Foundations of Arithmetic* (1884), he undertook the logical investigation of the fine structure of the concept of number. He rejected the Kantian view that arithmetical truths are synthetic **a priori,** whose knowledge involves intuition, and defended **Leibniz**'s insight that arithmetical truths are truths of reason and, in particular, truths of logic (this view came to be known as *logicism*). He argued against **Mill** that arithmetical laws are not empirical generalisations and against **Berkeley** that numbers are not subjective entities (e.g., ideas). He claimed that numbers are non-sensible, objective objects and his fundamental thought, as he put it, was that the content of a statement of number is an assertion about a **concept**. For instance, to say that the number of the moons of Venus is zero is to say that nothing falls under the concept 'moon of Venus' – which is to say that the number zero belongs to the concept 'moon of Venus'. He then tried to develop his theory of how numbers, *qua* **abstract entities**, are given to us (since they are given to us neither in experience nor in intuition). The key idea was that numbers are given to us via the **truth** of certain judgements, namely, numerical identities. Frege's overal approach was characterised by three principles that became very popular and controversial: (1) anti-psychologism: always to separate sharply the psychological from the logical, the subjective from the objective; (2) the context principle: never to ask the meaning of a word in isolation, but only in the context of a proposition; and (3) dualism: never to lose sight of the distinction between **concept** and object.

See **Abstraction principles; Analytic/synthetic distinction; Description theories of reference; Hilbert; Platonism, mathematical; Sense and reference**

Further reading: Frege (1884); Weiner (2004)

Function see **Functional explanation**

Functional explanation: It explains the presence of some item in a system in terms of the effects that this item has in the system of which it is a part. In biology, it is typical to explain a feature (a phenotypic characteristic) of a species in terms of its contribution to the enhancement of the chances of survival and reproduction. It is equally commonplace to explain the **properties** or the behaviour of the parts of an organism in terms of their functions in the whole: they contribute to the adequate functioning, the survival and reproduction of the whole. The **explanation** of the beating of the heart by appeal to its function to circulate the blood is a standard example of a functional explanation. Functional explanations are often characterised by the occurrence of teleological expressions such as 'the function of', 'the role of', 'in order to', 'for the purpose of'. It seems, then, that functional explanations explain the presence of an entity by reference to its effects. Hence, they seem to defy strict causal analysis. **Hempel** and **Nagel** tried to show how functional explanation could be understood in a way that had no serious teleological implications. One of the main problems was the presence of functional equivalents, that is, the existence of different ways to perform a certain function (for instance, artificial hearts might circulate the blood). Wouldn't it be proper, for instance, to explain the presence of heartbeat by claiming that it is a necessary **condition** for the proper working of the organism? We could argue thus: the presence of the heartbeat is a necessary condition for the proper working of the organism; the organism works properly; hence, the organism has a heart. The existence of functional equivalents shows that the intended conclusion does not follow. At best, all that could be inferred is the presence of one of the several items of a class of things capable of performing a certain function. Hempel thought that explanation in terms of functions works only in a limited sense and has only heuristic value. Faced with the same problem, Nagel suggested that under a sufficiently precise characterisation of the type of organism dealt with, only one kind of **mechanism** will be apt to fulfil the required function. Here is the form that functional explanations have (illustrated by Nagel's favourite example):

1. This plant performs photosynthesis.
2. Chlorophyll is a *necessary condition* for plants to perform photosynthesis.
3. Hence, this plant contains chlorophyll.

Any appearance of teleology in the functional explanation has gone. But, as Nagel stressed, this is *not* a causal explanation of the presence of chlorophyll. Functional explanation is then made to fit within the **deductive-nomological model**, but at the price of ceasing to be causal. There are two ways to react to Nagel's suggestion. One is to try to restore the causal character of functional explanation. The other is to deny that explanations have to be arguments. Both ways were put together in Larry Wright's (born 1937) *etiological* model of functional explanation. 'Etiology' means finding the causes. Etiological explanation is *causal* explanation: it concerns the causal background of the phenomenon under investigation. The basic pattern of functional explanation is:

The function of X is Z iff:

1. X is there because it does (results in) Z;
2. Z is a consequence (result) of X's being there.

For instance, the function of chlorophyll in plants is to perform photosynthesis if and only if chlorophyll is there because it performs photosynthesis and photosynthesis is a consequence of the presence of chlorophyll. An important feature of Wright's account is that it is suitable for explanation in biology, where the notion of natural selection looms large: natural (biological) functions are the results of natural selection because they have endowed their bearers with an evolutionary advantage. Consequently, etiological explanation does not reverse the causal order: a function is performed because it has been causally efficacious *in the past* in achieving a certain goal. According to Robert Cummins (born 1944), to ascribe a function to an item which is part of a system S is to ascribe to it some capacity in virtue of which it contributes to the capacities of the whole system S. So, functional explanations explain how a system can perform (i.e., has the capacity to perform) a certain complex task by reference to the capacities of the parts of the system to perform a series of subtasks that add up to the system's capacity.

See **Darwin; Explanation, causal**

Further reading: Hempel (1965); Nagel (1977); Wright (1976)

Galileo Galilei (1564–1642): Italian scientist and natural philosopher, one of the founders of modern science. He is the author of *Dialogue Concerning the Two Chief Systems of the World* (1632), in which he defended the Copernican heliocentric system against the Aristotelian cosmology, and *Discourse Concerning Two New Sciences* (1638), in which he laid the foundations of the new science of mechanics. Galileo famously argued that the book of nature is written in the language of mathematics. Though Galileo emphasised the role of experiment in science, he also drew a distinction between appearances and **reality**, which set the stage for explanatory theories of the phenomena that posited **unobservable entities**. The very possibility of the truth of **Copernicus**'s theory suggested that the world might not be the way it is revealed to us by the senses. The mathematical theories of motion he advanced were based on idealisations and abstractions. For Galileo, experience provides the raw material for these idealisations (frictionless inclined planes or ideal pendula), but the key element of the **scientific method** was the extraction, via **abstraction** and idealisation, of the basic structure of a phenomenon in virtue of which it can be translated into mathematical form. It is then that mathematical demonstration takes over and further consequences are deduced, which are tested empirically. Galileo also advanced a distinction between primary qualities and secondary ones. Primary are those qualities, like shape, size and motion, that are possessed by the objects in themselves, are immutable and objective and amenable to mathematical exploration. Secondary are those qualities, like colour and taste, that are relative, subjective and fleeting. They are caused on the senses by the primary qualities of objects, but, Galileo thought, in and of themselves, they are merely names. The world studied by science is a world of primary qualities: the subjective qualities can be left out of science without any loss.

See **Locke; Thought experiment**

Further reading: Galileo (1938)

Giere, Ronald (born 1938): American philosopher of science, author of *Explaining Science: A Cognitive Approach* (1988) and *Science without Laws* (1999). He has been one of the leading defenders of the **semantic view of theories**. He has also defended **methodological naturalism**. In his more recent work, he has argued that cognitive science should play the role of a general framework within which key philosophical issues about science are analysed and explained. He has denied that science needs, or should aim at, universal **laws of nature** and has defended a perspectival realism, according to which theories (understood as maps) offer us only perspectives on limited aspects of **reality**.

Further reading: Giere (1999)

Given, the: The supposed non-conceptual element of experience. Its existence was defended by many empiricists who were foundationalists. **Sense data** were supposed to be the immediate and indubitable contents of experience. They were supposed to act as the certain foundation of all **knowledge**. In his attack of the 'myth of the given', **Sellars** set up the following dilemma for foundationalist **empiricism**. The 'given' is either something with propositional content or not. If it does not have propositional content (if, say, it is an object or an **event**), it cannot confer any warrant on beliefs simply because, without any propositional content, it cannot function as a premise in a justificatory **argument**. If, in contrast, it does have propositional content, it cannot be justified independently of other things we know. For instance, the statement 'this is red' is not self-justified. Its being justified depends on its utterer knowing a host of other things, and in particular on her knowing a host of things about the reliability of observational reports. Hence, it cannot act as the type of certain foundation of knowledge, as the foundationalist demands.

See **Certainty; Foundationalism; Reliabilism**
Further reading: Sellars (1963)

Glymour, Clark (born 1942): American philosopher of science, author of *Theory and Evidence* (1980) and *Causation, Prediction and Search* (together with Peter Spirtes and Richard Scheines) (2000). He has worked in the philosophy of physics (especially, the philosophy of **space** and **time**), **confirmation** theory, **causation** and the philosophy of artificial intelligence. He advanced the quite influential **bootstrapping** account of confirmation and has been a critic of **Bayesianism**. Together with his collaborators, he did pioneering work on causal modelling and causal **inference**.

See **Old evidence, problem of**
Further reading: Glymour (1980)

Goodman, Nelson (1906–1998): American philosopher, author of *The Structure of Appearance* (1951), *Fact, Fiction,* *and Forecast* (1954) and *Ways of Worldmaking* (1978). Early in his career, he defended **nominalism** and tried to advance **Carnap**'s programme of the construction of the world out of a phenomenal basis. He is famous for the new riddle of **induction**, which shook up **Hempel**'s syntactic theory of **confirmation** and highlighted the need for natural **properties** in it. He offered a systematic analysis of the truth-conditions of **counterfactual conditionals** and defended the view that **laws of nature** are those generalisations that have privileged epistemic status in our cognitive inquiry (they are used in prediction and **explanation;** they are confirmed by their instances etc.). Later on in his career, he took a constructivist and relativist turn: the several symbolic systems (science, the arts etc.) are ways of worldmaking, that is, ways to construct the world.

See **Confirmation, Hempel's theory of; Grue**
Further reading: Goodman (1954)

Grue: Predicate introduced by **Goodman** in an attempt to pose a new riddle of **induction**. 'Grue' is defined as follows: observed before 2010 and found green, or not observed before 2010 and it is blue. All observed emeralds are green. But they are also *grue*. Why then should we take the relevant generalisation (or law) to be *All emeralds are green* instead of *All emeralds are grue*? Goodman argued that only the first statement ('All emeralds are green') is capable of expressing a law of nature because only this is confirmed by the observation of green emeralds. He disqualified the generalisation 'All emeralds are grue' on the grounds that the predicate 'is grue', unlike the predicate 'is green', does not pick out a **natural kind**. As he put it, the predicate 'is grue' is not projectable, that is, it cannot be legitimately applied (projected) to hitherto unexamined emeralds. Whether or not a generalisation will count as lawlike will depend on what kinds of predicates are involved in its expression.

See **Laws of nature**

Further reading: Stalker (1994)

Grünbaum, Adolf (born 1923): German-born American philosopher of science, founding director, in 1960, of the Centre for Philosophy of Science, University of Pittsburgh. He is the author of *Philosophical Problem of Space and Time* (1963) and *The Foundations of Psychoanalysis: A Philosophical Critique* (1984). He has worked on the philosophical foundations of relativity theory, defending a version of geometric **conventionalism** – in particular, the view that physical **space** does not possess intrinsic metric and that, hence, metric is imposed on it extrinsically. He also argued against **Popper**'s falsificationist criterion of **demarcation** of science from pseudo-science. For him, the important issue is not drawing a firm distinction between science and pseudo-science, but rather the cognitive accountability of science, that is, the procedures and methods that establish the epistemic credentials of scientific theories.

Further reading: Grünbaum (1973)

Hacking, Ian (born 1936): Canadian philosopher, one of the most influential philosophers of science of the second half of the twentieth century. He is the author of *Logic of Statistical Inference* (1965) and *Representing and Intervening* (1983). He has also written extensively on the history and philosophy of the concept of **probability** and, more recently, on **social constructivism**. He was one of the first contemporary philosophers who took **experiment**

seriously and emphasised that experimental practice has a life of its own, quite independently of theory-testing. He advanced **entity realism** and stressed the role of intervention in nature. His work on the concept of probability was based on the seminal idea, due to the historian of science A. C. Crombie (1915–1996), of a *style of reasoning*. Probabilistic thinking, according to Hacking, represented the emergence of a new style of reasoning, which was shaped around the novel concept of probability and its laws. Styles of reasoning introduce new objects, new types of evidence, new types of **argument** and **explanation**, and open up hitherto unexplored possibilities.

Further reading: Hacking (1965, 1983)

Hanson, Norwood Russell (1922–1967): American philosopher of science, author of *Patterns of Discovery* (1958). He was influenced by the later Wittgenstein, and, in his turn, deeply influenced **Kuhn** and **Feyerabend**. He relied on the Wittgensteinian idea that there is not a ready-made world. Rather, what there is and what one is committed to depends on the 'logical grammar' of the language one uses to speak of the world. For Hanson, science is a 'language game' characterised by its norms, rules, practices and concepts, but all these are *internal* to the game: they do not give the language users purchase on an independent world. He favoured **abduction** and thought that it led to fruitful hypotheses concerning the causes of observable phenomena. Hanson made possible a non-sceptical version of scientific **anti-realism**: science is not in the business of discovering the structure of a mind-independent world; rather, it is the language game which imposes structure onto the world and which specifies what facts there are.

Further reading: Hanson (1958)

Harding, Sandra (born 1935): American feminist philosopher of science, author of *The Science Question in Feminism*

(1986) and *Whose Science? Whose Knowledge?: Thinking from Women's Lives* (1991). She has advocated the **feminist standpoint**, arguing for a reconceptualisation of **objectivity** that starts from the lives of women and other marginalised subjects.

Further reading: Harding (1986)

Harré, Rom (born 1927): New Zealand-born philosopher of science who has spent most of his career in Oxford. He is the author of *Causal Powers* (1975, together with E. H. Madden) and *Varieties of Realism* (1986). He has also contributed to the philosophy of psychology. He was an early defender of a neo-Aristotelian (non-Humean) metaphysics and in particular of the view that properties are **powers** and things behave the way they do in virtue of their natures.

Further reading: Harré and Madden (1975)

Hempel, Carl Gustav (1905–1997): German-born American philosopher of science, with ground-breaking contributions to most areas of the philosophy of science, including the theory of meaning and **concept** formation, **explanation** and **confirmation**. He authored *Aspects of Scientific Explanation* (1965) which set the agenda for most subsequent thinking about explanation. Hempel was a member of the **Vienna Circle** and emigrated to the USA in 1937, where he taught at Princeton University and the University of Pittsburgh. In his early work, he downplayed the concept of **truth** and took the concepts of confirmation and **acceptance** as crucial tools for understanding the nature of epistemic commitment. Even quite late in his career, he claimed that the aim of scientific theorising is not truth but the optimal epistemic integration of the belief system that we hold at a given time. In the 1940s and 1950s he worked on the empiricist criterion of cognitive significance and the logic of confirmation. He was attracted to semantic holism, arguing that the meaning of a statement in a language is reflected in its logical relationships to all other statements in that language and not to the observational statements alone. He criticised **operationalism** and defended the view that theoretical concepts exhibit 'openness of content'. Eventually, he abandoned the distinction between observational terms and theoretical ones and spoke of 'antecedently understood' vocabulary. In the 1950s and 1960s he systematised the **deductive-nomological model of explanation** and advanced the **inductive-statistical model of explanation**. He also worked on **functional explanation**. He moved towards a stance more friendly to **scientific realism**, by criticising **Craig's theorem** and claiming that theories are indispensable in establishing an **inductive systematisation** of the phenomena.

See **Holism, semantic; Paradox of the ravens; Theoretician's dilemma**

Further reading: Hempel (1965)

Hertz, Heinrich (1857–1894): German physicist. His work on the foundations of mechanics led him to formulate the principles of mechanics in a new way, dispensing with the concept FORCE. His views were presented in the posthumously published book *The Principles of Mechanics Presented In a New Form* (1894). Based on the claim that forces acting at-a-distance were inconsistent with **Maxwell**'s electromagnetic theory, Hertz developed a system of mechanics founded solely on the concepts SPACE, TIME and MASS. Even though for Hertz the electromagnetic phenomena fell within the general province of mechanical phenomena, he took it to be premature to try to explain the laws of electromagnetism on the basis of the laws of mechanics. It was in this context that

Hertz uttered the now famous phrase 'Maxwell's theory is in Maxwell's system of equations'. Hertz claimed that theories are *images* or *pictures* in thought of things in the world and that the ultimate requirement for the admissibility of theories is this: the consequents of the images in thought are the images of the consequents of the things in nature. Though he thought that one image is more appropriate than another if it is simpler and it pictures more essential relations of the world, he insisted that there was no simple recipe for classifying theories (images) in terms of their appropriateness.

Further reading: Hertz (1894)

Hesse, Mary (born 1924): British philosopher of science, author of *Models and Analogies in Science* (1966) and *The Structure of Scientific Inference* (1974). She developed a theory for the role of **analogy** in science, based on extensive work on the concepts of force and field. She also developed a *network model* of scientific theories, which laid emphasis on the nomological interconnections of scientific **concepts** and denied any privileged nature to observational concepts.

Further reading: Hesse (1966)

Hilbert, David (1862–1943): German mathematician, one of the most famous mathematicians of all time. In *Foundations of Geometry* (1899), he showed that Euclid's five axioms were far from sufficient for the development of **Euclidean geometry**. Hilbert showed how Euclidean geometry could be cast in the form of a purely formal logical–mathematical axiomatic system, based on a new enlarged set of axioms. Hilbert's breakthrough, however, consists in his point that the deductive power of an axiomatic system is independent of the meaning of its terms and predicates and dependent only on their logical relationships. Hence, when it comes to what can be deduced from the axioms, the intuitive meanings of terms such as 'point', 'line', 'plane' etc. play no role at all. A Hilbert-style introduction of a set of terms by means of axioms is called implicit definition. Hilbert's approach to arithmetic has been characterised as formalism. Hilbert disagreed with **Frege** that mathematics is reducible to logic and agreed with **Kant** that mathematics has a specific subject matter. However, he took this subject matter to be not the form of intuition but rather a set of concrete extralogical objects, namely, symbols – numerals in the case of arithmetic. Hilbert thought that total infinities were illusions. But in order to accommodate the role that infinity plays in mathematics, he introduced ideal elements, along the lines of the ideal points at infinity in geometry. Given his thought that proving the consistency of a formal system is all that is required for using it, Hilbert took it that the search for **truth** should give its place to the search for consistency.

See **Definition, implicit; Syntactic view of theories**

Further reading: Hilbert (1899); Shapiro (1997)

Holism, confirmational: The view that theories are confirmed as wholes. Accordingly, when a theory is confirmed by the **evidence**, everything that the theory asserts or implies is confirmed. Confirmational holism, conjoined with the denial of the **analytic/synthetic distinction** and the thought that confirmable theories are meaningful, leads to semantic holism. In particular, it leads to the view that even mathematical or logical statements are confirmable by the evidence and hence that they have empirical content (i.e., they are not analytic truths).

See **Confirmation; Holism, semantic**

Further reading: Fodor and Lepore (1992); Quine (1951)

Holism, semantic: The view that all terms (or **concepts**) acquire their meaning from the theories and network of nomological statements in which they are embedded. This view became popular in the 1960s especially in connection with the meaning of **theoretical terms**. **Putnam** argued that all theoretical concepts are 'law-cluster' concepts: they get their meaning via the plethora of nomological statements in which they occur. Since these nomological statements are synthetic, there is no way to separate out those that fix the meaning of a concept and those that specify its empirical content. Hence, there is no way to draw the **analytic/synthetic distinction**. Semantic holism has contributed significantly to the wide acceptance of the claim that theoretical discourse is meaningful. However, combined with the thesis that all observation is theory laden, semantic holism can lead to the conclusion that the meaning of **observational terms** too is specified in a holistic way. Worse than that, since the meanings of terms is determined by the theory as a whole, it could now be claimed that every time the theory changes the meanings of *all* terms change. We have then a thesis of radical meaning variance in theory change. If, on top of that, it is also accepted that meaning determines reference (as the traditional **description theories of reference** have), an even more radical thesis follows, namely, reference variance. In order to avoid this consequence, some empiricists tried to hang onto the idea that observational terms are special: they do not get their meaning via theory. But semantic holism can be moderate. It may suggest that though terms do not get their meaning in isolation but within a network of lawlike statements and theories, *not* all parts of the network are inextricably interlocked in fixing the meaning of terms.

See **Causal theory of reference; Fodor; Incommensurability; Observation, theory-ladenness of**

Further reading: Carnap (1956); Fodor and Lepore (1992)

Hull, David (born 1935): American philosopher of science (and of biology, in particular), author of *Science as a Process* (1988). He has been a vocal critic of **essentialism,** arguing that biological species cannot be modelled on the basis of an essentialist metaphysics: a biological species does not possess essential properties, that is, properties the lack of which would make an individual not be a member of the species. He has also argued that biological species are individuals in that they evolve.

See **Evolutionary epistemology**
Further reading: Hull (1988)

Hume, David (1711–76): Scottish philosopher, author of the ground-breaking *A Treatise of Human Nature* (1739–40). In his *An Enquiry Concerning Human Understanding* (1748), Hume drew a sharp distinction between relations of ideas and matters of fact. Relations of ideas mark a special kind of truths, which are necessary and knowable a priori. Matters of fact, in contrast, capture contingent truths that are known a posteriori. This bifurcation leaves no space for a third category of synthetic **a priori** principles, the existence of which Hume firmly denied. Hume argued that all factual (and causal) knowledge stems from experience. He revolted against the traditional view that the necessity which links cause and effect is the same as the logical necessity of a demonstrative argument. He argued that there can be *no* a priori demonstration of any causal connection, since the cause can be conceived without its effect and conversely. Taking a cue from Malebranche, he argued that there was no impression of the supposed **necessary connection** between cause and effect. He also found inadequate, because circular, his predecessors' attempts to explain the link between causes and effects in terms of **powers,** active forces etc. His far-reaching point was that the alleged necessity of causal connection cannot be proved empirically. Any attempt to show, based on experience, that a regularity that has held in the past *will* or *must* continue to hold in the future will be circular and question-begging. It will presuppose a **principle of uniformity of nature.** But this principle is *not* a priori true. Nor can it be proved empirically without circularity. This Humean challenge to any attempt to establish the necessity of causal connections on empirical grounds has become known as his **scepticism** about **induction.** Hume never doubted that people think and reason inductively. He just took this to be a fundamental psychological fact about human beings which cannot be accommodated within the confines of the traditional conception of Reason. In his analysis of **causation,** Hume faced a puzzle. According to his empiricist theory of ideas, there are no ideas in the mind unless there were prior impressions. Yet, the concept of causation involves the idea of necessary connection. Since there is no impression of necessity in causal sequences, the *source* of this idea is the perception of a **constant conjunction** which leads the mind to form a certain habit or custom: to make a 'customary transition' from cause to effect. It is this felt determination of the mind that affords us the idea of necessity. So instead of ascribing the idea of necessity to a feature of the natural world, Hume took it to arise from *within* the human mind, when the latter is conditioned by the observation of a regularity in nature to form an expectation of the effect, when the cause is present. Hume claimed that the supposed objective necessity in nature is *spread* by the mind onto the world.

See **Kant; Laws of nature**

Further reading: Hume (1739); Stroud (1977)

Humean supervenience: The view that all causal facts supervene on non-causal facts. A standard way to cast this view is this: if two possible worlds are identical vis-à-vis their non-causal facts, they are also identical with respect to their causal facts. A chief advocate of this view was **Lewis,** who took it that if the spatio-temporal distribution of local qualities is fixed, then everything else, including facts about causal relations, is fixed.

See **Causation; Laws of nature; Supervenience**
Further reading: Loewer (1996)

Hypothetico-deductive method: It is based on the following idea. Form a hypothesis H and derive some observational consequences from it. If the consequences are borne out, the hypothesis is confirmed (accepted). If they are not borne out, the hypothesis is disconfirmed (rejected). To be sure, the observational consequences follow from the conjunction of H with some statements of initial conditions, other auxiliary assumptions and some bridge-principles which connect the vocabulary in which H is couched and the vocabulary in which the observational consequences are couched. It is these bridge-principles that make the hypothetico-deductive method quite powerful, since they allow for what may be called 'vertical extrapolation' – to be contrasted with the 'horizontal extrapolation' characteristic of **enumerative induction.** There are two main problems that plague the hypothetico-deductive method. The first is a version of the **Duhem–Quine** problem. Since it is typically the case that, in applications of the hypothetico-deductive method, the predictions follow from the conjunction of the hypothesis with other auxiliary assumptions and initial and boundary conditions, when the prediction is *not* borne out it is the whole cluster of premises that gets refuted. But the hypothetico-deductive method alone cannot tell us how to apportion praise and blame among them. At least one of them is false but the culprit is not specified by the hypothetico-deductive method. It might be that the hypothesis is wrong, or some of the auxiliaries were inappropriate. In order to pinpoint the culprit, we need further information, namely, whether the hypothesis is warranted enough to be held on to, or whether the auxiliaries are vulnerable to substantive criticism etc. But all these considerations go far beyond the deductive link between hypotheses and evidence that forms the backbone of the hypothetico-deductive method and are not incorporated into its logical structure. The other problem faced by the hypothetico-deductive method may be called the problem of alternative hypotheses: there may be other hypotheses that entail the very same predictions. If the warrant for H is solely based on the fact that it entails the evidence, insofar as there is another hypothesis H^* which also entails the **evidence**, H and H^* will be equally warranted. Hence, the hypothetico-deductive method will offer no way to discriminate between mutually incompatible but empirically equivalent hypotheses in terms of warrant.

See **Confirmation; Descartes; Tacking paradox, the** Further reading: Gower (1998); Salmon (1967)

Idealisation see **Abstraction**

Idealism: The view that everything that exists is either a mind or dependent on minds. It has been associated with **Berkeley** who argued that unthinking things are collections of ideas and ideas exist in so far as they are being perceived. By tying existence to perceiving minds (and, ultimately, God himself), idealism was meant to block

scepticism. Idealism does not deny that ordinary things like tables and chairs, or even more exotic things like electrons and quarks, exist – rather it asserts that their existence is mind-dependent. It is opposed to realism, which argues that unthinking things are mind-independent. A common argument against idealism is that it runs together the *act* of perceiving (which involves the mind) with the *object* of perception (which might well be something mind-independent).

See **Realism and anti-realism**

Further reading: Stove (1991)

Incommensurability: Term introduced by **Feyerabend** and **Kuhn** to capture the relation between **paradigms** before and after a scientific revolution. The pre-revolutionary and the post-revolutionary paradigms were said to be incommensurable in that there was no strict translation of the terms and predicates of the old paradigm into those of the new. Though Kuhn developed this notion in several distinct ways, its core is captured by the thought that two theories are incommensurable if there is no language into which both theories can be translated without residue or loss. Kuhn supplemented this notion of untranslatability with the notion of *lexical structure*: two theories are incommensurable if their lexical structures (i.e., their taxonomies of **natural kinds**) cannot be mapped into each other. When competing paradigms have locally different lexical structures, their incommensurability is local rather than global.

See **Holism, semantic; Observation, theory-ladenness of**

Further reading: Kuhn (1962); Sankey (1994)

Induction see **Eliminative induction; Enumerative induction; Induction, the problem of; Inductive logic; Laplace**

Induction, new riddle of see **Grue**

Induction, the problem of: The problem of justifying the **infer-ence** from the observed to the unobserved; or from par-ticular instances to generalisations; or from the past to the future. It has been particularly acute for nominalists, who deny the existence of **universals**. Realists about uni-versals thought they could justify induction: after a survey of a relatively limited number of instances, the thought ascended to the universal (what is shared in common by these instances) and thus arrived at truths which were general, necessary and unrevisable. This kind of route was closed for nominalists. They had to rely on experi-ence through and through and inductive generalisations based on experience could not yield certain and neces-sary **knowledge**. The problem of the rational grounds for induction came into sharp focus in **Hume**'s work. His **scepticism** about induction is the claim that any attempt to show, based on experience, that a regularity that has held in the past *will* or *must* continue to hold in the fu-ture will be circular and question-begging. **Mill**, a radical inductivist, never thought there was a *problem* of induc-tion. He took it that induction did not need any justifi-cation. The justification of induction started to become a problem in the late nineteenth and the early twentieth centuries. John Venn (1834–1923) took it to be the prob-lem of establishing the foundation of the belief in the uniformity of nature and argued that this belief should be taken as a logical postulate, while the issue of its ori-gin should be relegated to psychology. It was John May-nard Keynes (1883–1946), in his *Treatise on Probability* (1921), who first interpreted Hume's critique of **causa-tion** as being about inductive reasoning. Keynes, and fol-lowing him **Carnap**, tried to solve the problem of induc-tion by turning induction into a kind of logic – **inductive**

logic – operating on the basis of the laws of **probability** and logical or quasi-logical principles (such as the **principle of indifference**). **Reichenbach** aimed at a pragmatic vindication of induction. He argued that though the **principle of uniformity of nature** cannot be proved or established empirically, *if* nature is uniform, induction (and, in particular, the **straight rule of induction**) will succeed, in the long run, in uncovering the regularities that exist in the phenomena. *If*, in contrast, induction fails, any other method will fail. Currently, the problem of induction has been set within the subjective Bayesian framework. The key thought is that agents start with their subjective prior degrees of belief and then update them by **conditionalisation**. Induction, then, is the process of updating already possessed **degrees of belief** and its **justification** gives way to the problem of justifying conditionalisation on the **evidence**. Popperians deny that there is any problem of induction, since they deny that there is such a thing as induction. **Naturalism** denies a presupposition that all those who have tried to justify induction have shared, namely, that induction needs justification and that a method cannot be relied upon unless it is first justified on independent grounds. Naturalists argue that insofar as induction is reliable (and no one has shown that it is not) it can and does lead to warranted beliefs.

See **Bayesianism; Confirmation; Corroboration; Nominalism; Principle of induction; Reliabilism; Validation vs vindication**

Further reading: Howson (2000); Kneale (1949); Swinburne (1974)

Inductive logic: A formal system based on the **probability** calculus that aims to capture in a logical and quantitative way the notion of inductive support that **evidence** accrues to a hypothesis or theory. Being a logic, this system mimics the content-insensitive structure of deductive logic. It was advanced by Keynes and was developed into a rigorous system by **Carnap**. The key idea is that **confirmation** is a logical relation between statements, those that express the evidence and those that express the hypothesis. This logical relation is called the degree of **partial entailment** of a hypothesis by the observational evidence. Carnap tried to devise certain quantitative functions that capture statements of the form: the degree of confirmation of H by e is r, where r is a real number between 0 and 1. Carnapian inductive logic was meant to be justified **a priori**. Hence, he relied on the **principle of indifference** to assign initial (prior) probabilities. But different applications of this principle lead to inconsistent results. One can apply the principle of indifference to state-descriptions. These are complete ways the world might be. Given a formal language L with constants and predicates, a state-description is a conjunction of sentences which describe completely a possible state of the domain of individuals of L with respect to all attributes (i.e., properties and relations). But it turned out that the resulting confirmation-function (what Carnap called $c^{\dagger}$) did not allow for learning from experience. No evidence could raise the (posterior) probability of a state-description to more than what it was before the evidence rolled in. Alternatively, one might apply the principle of indifference to structure-descriptions. These are disjunctions of state-descriptions. The structure-description that corresponds to a class of state-descriptions is the disjunction of all the isomorphic state-descriptions, that is, of all state-descriptions that have the same **structure**. (Isomorphic state-descriptions differ only with respect to the names attached to their individuals.) The resulting confirmation-function (what Carnap called c^{*}) does allow for learning from experience, but at the price of putting a premium (higher prior probability) on some structure-descriptions, namely, those which assert that certain universal regularities are present in the world. Hence, it was no longer the case that this relation of confirmation was a logical relation, independent of substantive assumptions as to how the world is likely to be. Finally, Carnap devised the continuum of inductive methods, and drew the conclusion that there can be a variety of actual inductive methods whose results and effectiveness vary in accordance to how one picks out the value of a certain parameter, where this parameter depends on formal features of the language used. But there is no a priori reason to select a particular value of the relevant parameter, and hence there is no **explication** of inductive **inference** in a unique way.

See **Induction, the problem of**

Further reading: Carnap (1950b); Salmon (1967)

Inductive-statistical model of explanation: Developed by **Hempel** for the **explanation** of singular events whose **probability** of happening is less than unity. Suppose that Jones has suffered from a septic sore throat, which is an acute infection caused by bacteria known as *streptococcus hemolyticus*. He takes penicillin and recovers. There is no strict (deterministic) law that says that whoever is infected by *streptococcus* and takes penicillin will recover quickly. Hence, we cannot apply the **deductive-nomological model** to account for Jones's recovery. Suppose, however, that there is a statistical generalisation of the following form: whoever is infected by *streptococcus* and takes penicillin has a very high probability of recovery. Let's express this as follows: $\text{prob}(R/P \ \& \ S)$ is very high, where 'R' stands for quick recovery, 'P' stands for taking penicillin and 'S' stands for being infected by *streptococcus* germs. Given this statistical generalisation, and given that Jones was infected by *streptococcus* and took penicillin, the probability of Jones's quick recovery was high. We have *inductive* grounds to expect that Jones will recover. We can then construct an inductive **argument** that constitutes the basis of the explanation of an event whose occurrence is governed by a statistical generalisation. Generally, the logical form of an inductive-statistical explanation is this:

$$Fa$$
$$\text{prob}(G/F) = r, \text{ where r is high (close to 1)}$$
$$\overline{\overline{}} [r]$$
$$Ga.$$

The double line before the conclusion indicates that it is an *inductive* argument: the conclusion follows from the premises with high probability. The strength *r* of the inductive support that the premises lend to the conclusion is indicated in square brackets. The requirement of high probability is essential to the inductive-statistical model. Yet, it is clear that improbable events do occur and are in need of explanation. Besides, high probability is not sufficient for a good statistical explanation. Suppose we explain why Jones recovered from a common cold within a week by saying that he took a large quantity of vitamin C. We can construct an inductive-statistical argument, one of the premises of which is the following statistical law: the probability of recovery from common colds within a week, given taking vitamin C, is very high. Though the formal conditions for an inductive-statistical explanation are met, the inductive argument offered is not a good explanation of Jones's recovery: the statistical law is *irrelevant* to the explanation of recovery since common colds go away after a week, anyway. For the statistical law to be explanatory it should capture a causal law. In any case, as Hempel noted, the inductive-statistical model faces the problem of ambiguity. Briefly put, the problem is to what reference class to include the event to be explained (explanandum). Given that the explanandum may belong to lots of difference reference classes, which one shall we choose to specify the probability of the conclusion in the relevant inductive-statistical argument? Different choices of reference classes will lead to inductive-statistical arguments that have mutually consistent premises and yet incompatible conclusions. Hempel tried to mitigate this problem by introducing the further requirement of maximal specificity, which, in effect, takes the relevant reference class to be the narrowest class consistent with the total evidence that is available.

See **Probability, inductive; Statistical-relevance model of explanation**

Further reading: Hempel (1965); Psillos (2002); Salmon (1989)

Inductive systematisation: A theory is said to offer an inductive systematisation of a set of phenomena (empirical laws) if it establishes inductive connections among them, that is, if it can be used as a premise in inductive **arguments** whose other premises concern observable phenomena and whose conclusions refer to observable phenomena. Take, for instance, a hypothesis H (or a cluster thereof) that entails observational consequences O_1, $O_2 \ldots O_n$. When these obtain, although we cannot deductively infer H, we can inductively conclude that H holds. Suppose, further, that H together with other theoretical and observational hypotheses entail an extra testable prediction O_{n+1}. This new prediction could not have been issued by the observational consequences O_1, $O_2 \ldots O_n$ on their own. Its derivation rests essentially on accepting the inductively inferred hypothesis H. So, H is indispensable in establishing this inductive connection between $O_1, O_2 \ldots O_n$ and O_{n+1}. This idea was used by **Hempel** and others in order to block the instrumentalist claim that theories, seen as establishing solely deductive connections among observables, are dispensable.

See **Craig's theorem; Hypothetico-deductive method; Instrumentalism; Probability, inductive**

Further reading: Hempel (1965); Psillos (1999)

Inductivism: The view that **induction**, and **enumerative induction** in particular, is the ultimate basis of **knowledge**. It has been advocated by **Mill**.

Inference: A cognitive process in virtue of which a conclusion is drawn from a set of premises. It is meant to capture both the psychological process of drawing conclusions and the logical or formal rules that entitle (or justify) the subject to draw conclusions from certain premises. Inferences proceed via inferential rules (**argument** patterns). They can be divided into deductive (or demonstrative) and non-deductive (non-demonstrative or ampliative).

See **Ampliative inference; Deductive arguments; Inference to the best explanation; Probability, inductive**

Further reading: Harman (1986)

Inference to the best explanation: Mode of **inference** akin to Peircean **abduction**. The expression 'inference to the best explanation' was introduced by Gilbert Harman (born 1938), to capture the inferential process in which from the fact that a certain hypothesis, if true, would explain the **evidence**, an agent is entitled to infer the **truth** of that hypothesis. The warrant for the acceptance of a hypothesis is based on the explanatory quality of this hypothesis, on its own but also taken in comparison with others. Given that, as a rule, there will be several other hypotheses that explain the evidence, an agent must have grounds to reject all such alternative hypotheses before she is warranted in making the inference. Explanatory power is connected with the basic function of an **explanation**, namely, providing understanding. The evaluation of explanatory power takes place in two directions. The *first* is to look at the specific background information (beliefs) that operate in a certain application of inference to the best explanation. The *second* is to look at a number of structural features (standards) which competing explanations might possess. Candidates for such standards are: completeness, **simplicity**, **unification** and precision. However, though many philosophers grant that these standards have some genuine connection with explanatory quality or merit, they question their epistemic status: why are they anything more than pragmatic virtues? Others claim that these virtues possess a straight cognitive function: they safeguard the explanatory coherence of our total body of belief as well as the coherence between our body of belief and a new potential explanation of the evidence.

See **Coherentism**; **Eliminative induction**; **No-miracles argument**

Further reading: Lipton (2004)

Innate ideas see **Concept empiricism**; **Rationalism**

Instrumentalism: View about science according to which theories should be seen as (useful) instruments for the organisation, classification and prediction of observable phenomena. The 'cash value' of **scientific theories** is fully captured by what theories say about the observable world. Instrumentalism comes in different forms: syntactic and semantic. *Syntactic* instrumentalism treats the theoretical claims of theories as syntactic-mathematical constructs which lack truth-conditions, and hence any assertoric content. It comes in two varieties: eliminative and non-eliminative. The non-eliminative variety (associated with **Duhem**) takes it that one *need not* assume the existence of an unobservable **reality** behind the phenomena, nor that science aims to describe it, in order to do science and to do it successfully. Eliminative instrumentalism takes a stronger view: theories should not aim to represent anything 'deeper' than experience, because, ultimately, there is nothing deeper than experience (an unobservable reality) for the theories to represent. Faced with the challenge that theoretical assertions seem meaningful and aim to describe an unobservable **reality**, eliminative instrumentalists have appealed to **Craig's theorem** to defend the view that the theoretical vocabulary is eliminable *en masse* and hence that the question of whether they can refer to **unobservable entities** is not even raised. *Semantic* instrumentalism takes theoretical statements to be meaningful but only in so far as (and because) they are fully translatable into assertions involving only observational terms. If theoretical statements are fully translatable, they end up being nothing but disguised talk about observables, and hence they are ontologically innocuous: they should not be taken to refer to unobservable entities, and hence they license no commitments to them. The prime problem of syntactic instrumentalism is that it fails to explain how scientific theories can be empirically successful, especially when it comes to **novel predictions**. If theories fail to describe (even approximately) an unobservable reality, it is hard to explain why theories can be, as Duhem put it, 'prophets for us'. The prime problem with semantic instrumentalism is that theories have excess content over their observational consequences in that what they assert cannot be fully captured by what theories say about the observable phenomena. It is noteworthy that attempts to translate **theoretical terms** into observational ones have all patently failed.

See **Reductive empiricism; Scientific realism**

Further reading: Newton-Smith (1981); Psillos (1999)

Internal realism see **Putnam; Realism and anti-realism**

Intersubjective see **Objectivity**

Intrinsic vs extrinsic: Important distinction in the metaphysics of **properties** and relations. Those properties of an object are intrinsic that are compatible with loneliness, that is, those properties that an object would have even if it was the only object in the universe, for example, the shape of an object. Those properties are extrinsic that an object possesses in virtue of its being related to other objects, namely, the property of a book of being owned by Karl Marx. A (dyadic) relation is intrinsic to its *relata* if the following holds: when two *relata* stand in this relation, this is entirely a matter of how the two *relata* are vis-à-vis one another, and not at all a matter of their relations to other things. For instance, the relation *x has more mass than y* is an intrinsic relation of the pair <sun, earth>, since it is true that the sun has more mass than the earth and this depends entirely on how the sun and the earth are related to each other. By contrast, the relation that two objects *x* and *y* have when they belong to the same owner is extrinsic to these objects in the sense that whether it is true that *x* and *y* belong to the same owner will depend on their relation to a third thing (namely, the owner).

Further reading: Langton and Lewis (1998)

Inus conditions: Version of the regularity view of **causation** developed by John L. Mackie (1917–1981). Mackie stressed that effects have, typically, a plurality of causes. A certain effect can be brought about by a number of distinct clusters of factors. Each cluster is sufficient to bring about the effect, but none of them is necessary. The regularities in nature have a complex form ($A \, \& B \, \& \, C$ or $D \, \& E \, \& \, F$ or $G \, \& H \, \& \, I$) $\longleftrightarrow E$, which should be read as: all $(A \& B \& C$ or $D \& E \& F$ or $G \& H \& I)$ are followed by E, and all E are preceded by $(A \& B \& C$ or $D \& E \& F$ or $G \& H \& I)$. How do we pick out the cause of an event in this setting? Each single factor of the cluster $A \& B \& C$ (e.g., A) is related to the effect E in an important way. It is an *insufficient* but *non-redundant* part of an *unnecessary* but *sufficient* condition for E. Using the first letters of the italicised words, Mackie has called such a factor an *inus* condition. Causes, then, are *inus* conditions. So, to say that short circuits cause house fires is to say that the short circuit is an *inus* condition for house fires. It is an insufficient part because it cannot cause the fire on its own (other conditions such as oxygen, inflammable material etc. should be present). It is a non-redundant part because, without it, the rest of the conditions are not sufficient for the fire. It is just a part, and not the whole, of a sufficient condition (which includes oxygen, the presence of inflammable material etc.), but this whole sufficient condition is not necessary, since some other cluster of conditions, for example, an arsonist with petrol etc. can produce the fire.

See **Condition, necessary; Condition, sufficient**

Further reading: Mackie (1974)

Isomorphism see **Structure**

James, William (1842–1910): American philosopher and psychologist, one of the founders of **pragmatism**. He authored *The Principles of Psychology* (1890), which made him famous as a psychologist. In his *Pragmatism: A New Name for Some Old Ways of Thinking* (1907), he advanced the pragmatic method as a criterion for resolving disputes (especially metaphysical ones). One is supposed to ask: what difference in experience would the adoption of this or that view make? He took it that theories are instruments useful for practical purposes (especially the anticipation of nature) and not 'answers to enigmas' as he put it. On **truth**, he took the line that the truth lies in the process of verification of a proposition. Truth, as he put it, is *made* in the course of experience. In his *Will to believe* (1896), he argued for the ineliminable role of volition (will) in **belief**. James noted that in forming our opinion we pursue two main aims: *we must know the truth*; and *we must avoid error*. If we want to move between credulity and **scepticism** in forming our beliefs, we should strike a balance between the two aims. This calls for a *value judgement*, which is *not* an objective matter (even though achieving truth and avoiding error are). The appeal to the will is meant to capture this part of our doxastic (belief-forming) practices that goes beyond the demand for reasons and **evidence**.

See **Pascal's wager; Verificationism; Voluntarism**
Further reading: James (1897)

Judgement empiricism: The view that all judgements (statements, beliefs) must get their **justification** from experience. Hence, the justification for a **belief** must be either directly given in experience (e.g., by means of perception) or be a function of other beliefs whose own justification stems directly from experience. The radical version of this view takes even logical and mathematical beliefs to be justifiable empirically. More moderate versions allow for a certain kind of statements (analytic or meaning-fixing statements) to get their justification independently of experience.

See **Concept empiricism; Empiricism; Foundationalism**
Further reading: Reichenbach (1951); Russell (1912); Sellars (1963)

Justification: The property of a true **belief** that converts it to **knowledge**. Relatedly, the process of rendering a belief warranted. Beliefs can be justified, though false. Hence, justification has to do with what the subject does to secure his/her beliefs from error, even though he/she does not always succeed in this endeavour. According to externalist approaches to epistemology, justification is a state a subject is in if the subject has followed reliable methods of inquiry, or if his/her beliefs are caused in the right way, irrespective of whether he/she has reasons to support his/her beliefs or to consider reliable the methods followed. There have been many theories of justification, such as **foundationalism, coherentism** and **reliabilism**. In the philosophy of science, justification has been linked to **confirmation**. Relatedly, the literature in the philosophy of science has mostly focused on the justification of **scientific method,** and **induction** in particular. Recently, there have been contextualist approaches to justification, urging that what counts as justification may well vary from context to context.

Further reading: Plantinga (1993)

Kant, Immanuel (1724–1804): German philosopher, author of the ground-breaking *Critique of Pure Reason* (1781 – 2nd edn 1787). As he famously stated, it was **Hume**'s critique of necessity in nature that awoke Kant from his dogmatic slumber. Kant rejected strict **empiricism** and uncritical **rationalism**. He claimed that although all **knowledge** starts with experience it does not arise from it: it is actively shaped by the categories of the understanding and the forms of pure intuition (**space** and **time**). The mind imposes some conceptual and formal structure onto the world, without which no experience could be possible. Kant, however, thought there could be no knowledge of things as they were in themselves (*noumena*) and only knowledge of things as they appeared to us (*phenomena*). Be that as it may, his master thought was that some synthetic **a priori** principles should be in place for experience to be possible. These constitute the object of knowledge in general. In his three *Analogies of Experience*, Kant tried to prove that three general principles hold for all objects of experience: that substance is permanent; that all changes take place in conformity with the law of cause and effect; that all substances are in thoroughgoing interaction. These are synthetic a priori principles that make experience possible. Kant took the principle of **causation**, namely, that everything that happens presupposes something which it follows by rule, to be required for the mind to make sense of the temporal irreversibility that there is in certain sequences of impressions. Kant's aim in *Metaphysical Foundations of Natural Science* (1786) was to show how the transcendental principles of the understanding could be made concrete in the form of laws of matter in motion. These were metaphysical laws in that they determined the possible behaviour of matter in accordance with mathematical rules. Kant thus enunciated the law of conservation of the quantity of matter, the law of inertia and the law of equality of action and reaction and thought that these laws were the mechanical analogues (cases *in concreto*) of his general transcendental principles. They determine the pure and formal structure of motion, where motion is treated purely mathematically *in abstracto*. It is no accident that the last two of these laws are akin to Newton's law and that the first law was presupposed by **Newton** too. Kant's metaphysical foundations of (the possibility of) matter in motion were precisely meant to show how Newtonian mechanics was possible. Kant thought there could not be proper science without metaphysics. But he also thought that there are physical laws that are discovered empirically. Though philosophically impeccable, Kant's architectonic suffered severe blows in the nineteenth and the early twentieth centuries, coming mostly from developments in the sciences.

See **Analytic/synthetic distinction; A priori/a posteriori; Euclidean geometry; Neo-Kantianism; Non-Euclidean geometries**

Further reading: Guyer (1992); Kant (1787)

Knowledge: Justified true **belief**. This tripartite analysis, which goes back to Plato, has been the subject of great debate ever since 1963, when Edmund Gettier (born 1927) published some well-known counterexamples to it. These have aimed to establish that having a justified true belief is not sufficient for knowledge. There have been a number of theories trying either to supplement the traditional account or to reform it. Prominent among them has been the causal theory of knowledge, according to which knowledge is the state a subject is in if his/her true belief has been the effect of a causal chain which terminates in the fact that he/she knows.

See **Certainty; Justification; Truth**

Further reading: Pollock (1986)

Kripke, Saul (born 1940): American philosopher and logician, famous for his ground-breaking work in modal logic and the philosophy of language. In *Naming and Necessity* (1972), Kripke broke with the Kantian tradition which equated the necessary truths and the a priori truths as well as with the empiricist tradition which equated the necessary truths with the analytic truths. He argued that there are necessarily true statements that can be known a posteriori. Therefore, he made it possible to think of the existence of necessity in nature which is not quite the same as logical necessity, and yet strong enough to warrant the label 'necessity'. He also argued that there are contingent truths that are known a priori. So it is one thing to ask how the **truth** of a statement is known (a priori-a posteriori) and quite another to ask whether this truth could be otherwise (necessary-contingent). Kripke criticised the **description theories of reference** and advanced the causal-historical account of reference. Kripke based his views on necessity on an essentialist metaphysics, founded on a distinction between essential and accidental properties.

See **A priori/a posteriori; Causal theory of reference; Essentialism; Natural kinds**

Further reading: Fitch (2004); Kripke (1980)

Kuhn, Thomas (1922–1996): One of the most famous historians and philosophers of science of the twentieth century, author of *The Structure of Scientific Revolutions* (1962). His books include: *The Copernican Revolution* (1957) and *The Essential Tension* (1977). He was one of the architects of the historical turn of the 1960s. Kuhn's theory of science should be seen as the outcome of two inputs: (1) a reflection on the actual scientific practice as well as the actual historical development and succession of **scientific theories**; and (2) a reaction to what was perceived to be the dominant logical empiricist and Popperian images of scientific growth: a progressive and cumulative process that is governed by specific rules as to how **evidence** relates to theory. According to Kuhn, the emergence of a scientific discipline is characterised by the adoption by a community of a **paradigm**. A long period of normal science emerges, in which scientists attempt to apply, develop and explore the paradigm. During normal science, the paradigm is not under test or scrutiny. It is developed by means of an activity akin to puzzle solving, in that scientists follow the rules (or the concrete exemplars) laid out by the paradigm in order to (1) characterise which problems are solvable and (2) solve them. This rule-bound (or, better, exemplar-bound) activity that characterises normal science goes on until an anomaly appears. The emergence of anomalies signifies a decline in the puzzle-solving efficacy of the paradigm. The community enters a stage of crisis that is ultimately resolved by a revolutionary transition from the old paradigm to a new one. The new paradigm employs a different conceptual framework, and sets new problems as well as rules for their solutions. A new period of normal science emerges. Crucially, for Kuhn the change of paradigm is not rule governed. It has nothing to do with degrees of **confirmation** or conclusive refutations. Nor does it amount to a slow transition from one paradigm to the other. Rather, it is an abrupt change in which the new paradigm completely replaces the old one. Kuhn's philosophy can be seen as a version of **neo-Kantianism** because it implied a distinction between the world-in-itself, which is epistemically inaccessible to inquirers, and the phenomenal world, which is constituted by the concepts and categories of the inquirers, and is therefore epistemically accessible to them. But Kuhn's neo-Kantianism was *relativised*: he thought there was a plurality of phenomenal worlds, each being dependent on, or constituted by some community's paradigm. The paradigm imposes, so to speak, a structure on the world of appearances: it carves up this world in 'natural kinds'. But different paradigms carve up the world of appearances in different networks of **natural kinds**. **Incommensurability** then follows since it is claimed that there are no ways fully to match up the natural-kind structure of one paradigm with that of another's. Kuhn argued that there are some important traits that characterise a good scientific theory: accuracy, consistency, broad scope, **simplicity** and fruitfulness.

See **Holism, semantic; Observation, theory-ladenness of**

Further reading: Bird (2000); Kuhn (1962)

L

Lakatos, Imre (1922–1974): Hungarian-born philosopher who taught at the London School of Economics. He aimed to combine **Popper**'s and **Kuhn**'s images of science into a single model of theory-change, which preserves progress and rationality while it avoids Popper's naive **falsificationism** and respects the actual history of radical conceptual change in science. He developed the methodology of scientific research programmes. A research programme is a sequence of theory and is characterised by the hard core, the negative heuristic and the positive heuristic. The hard core comprises all those theoretical hypotheses that any theory which belongs to the research programme must share. The advocates of the research programme hold these hypotheses immune to revision. This methodological decision to protect the hard core constitutes the negative heuristic. The process of articulating directives as to how the research programme will be developed, either in the face of anomalies or in an attempt to cover new phenomena, constitutes the positive heuristic. It creates a protective belt around the hard core, which absorbs all potential blows from anomalies. A research programme is progressive as long as it yields **novel predictions**, some of which are corroborated. It becomes degenerative when it offers only *post hoc* accommodation of facts, either discovered by chance or predicted by a rival research programme. Progress in science occurs when a progressive research programme supersedes a degenerative one. Lakatos's methodology is retroactive: it provides no way to tell which of two currently competing research programmes is progressive. For, even if one of them seems to be stagnant, it may stage an impressive comeback in the future.

See **Ad hocness/Ad hoc hypotheses; Prediction vs accommodation**

Further reading: Lakatos (1970)

Laplace, Pierre Simon, Marquis de (1749–1827): French mathematician and astronomer, perhaps the main Newtoniam figure in France. His monumental five-volume *Celestial Mechanics*, which appeared between 1799 and 1825, extended and developed **Newton**'s gravitational theory. He developed the classical interpretation of **probability** in *A Philosophical Essay on Probabilities*. One of his chief claims was that nearly all **knowledge** is uncertain and that **induction** (as well as **analogy**) was based on probabilities. He thought, however, that probability is a measure of ignorance. His endorsement of **determinism** implied that a powerful mind (a vast intelligence, as he put it) who knew the **laws of nature** and the initial conditions of all bodies would have certain knowledge of the past as well as future events – thereby dispensing with probabilities. He devised a rule of induction, known as Laplace's rule of succession, according to which if the actual relative frequency of observed As that are Bs is m/n, our degree of confidence (i.e., the probability) that the next A will be B should be $m + 1/n + 2$. If, in particular, $m = n$ (if, that is, *all* observed As have been Bs), the probability that the next A will be B is: $n + 1/n + 2$, which clearly goes to unity, as n tends to infinity. Based on this rule, he was able to claim that the probability that the sun will rise tomorrow (given that it has risen every day for at least 5,000 years) is almost unity. As Laplace put it, it is a bet of 1,826,214 to 1 that the sun will rise tomorrow.

See **Induction, the problem of; Probability, classical interpretation of**

Further reading: Laplace (1814)

Laudan, Lawrence (born 1941): American philosopher of science, author of *Science and Values* (1984) and *Beyond Positivism and Relativism* (1996). He has been a vocal critic of **scientific realism** – advancing the argument from the **pessimistic induction**. However, he has also been one of the most severe critics of the argument from the **underdetermination of theories by evidence**. He has claimed that **evidence** that is not entailed by a theory can support it nonetheless; and, conversely, evidence entailed by the theory may not lend support to it. He has defended **pragmatism,** as an alternative to scientific realism, and has taken the instrumental reliability of science to be its distinctive characteristic. He has also defended normative **naturalism** – though he has denied that **truth** is or should be the aim of science, since, as he claimed, it would be a utopian aim. Laudan advanced the reticulational model of scientific **rationality**, according to which methods and aims of science change over time no less than theories do, but not all at the same time. New theories might introduce and legitimise new methods, and new methods might advance new aims or discredit old ones.

Further reading: Laudan (1996)

Laws of nature: Principles that govern the workings of nature. Understanding what laws of nature are has turned out to be a central problem in the philosophy of science. This is because laws have been implicated in **causation** and

explanation. An important tradition in the philosophy of science has been that causal explanation proceeds by subsuming the events to be explained under general laws; causation was intertwined with the presence of laws and explanation was taken to consist in a law-based demonstration of the explanandum. But though many endorse the centrality of laws in causation and explanation, there has been considerable disagreement as to what laws of nature are.

The regularity view: most empiricists adopted the *Regularity View of Laws:* laws are cosmic regularities. According to the Humean tradition, there are only regularities in nature, that is, sequences of event-types, which happen in **constant conjunction:** whenever one occurs, it is invariably followed by the other. When, for instance, it is said that it is a law that metals expand when heated (under constant pressure), Humeans mean that there is a regularity in nature according to which whenever a metal gets heated it expands. There is no necessity in this regularity because (1) it is logically possible that a metal is heated (under constant pressure) and yet it does not expand; and (2) there is nothing in the nature of a metal that makes it the case that, necessarily, it will expand when it is heated. Yet, empiricists have had a hurdle to jump: not all regularities are causal. Nor can all regularities be deemed laws of nature. So they were forced to draw a distinction between those regularities that constitute the laws of nature and those that are, as **Mill** put it, 'conjunctions in some sense accidental'. The predicament that Humeans are caught in is this. Something (let's call it the property of lawlikeness) must be added to a regularity to make it a law of nature. But what can this be?

The epistemic view: the first systematic attempt to characterise this elusive property of lawlikeness was broadly epistemic. The thought, advanced by Ayer, Richard Bevan

Braithwaite (1990–90) and **Goodman** among others, was that lawlikeness was a feature of those generalisations that play a certain *epistemic* role: they are believed to be true, and they are so believed because they are confirmed by their instances and are used in proper inductive reasoning. On this view, it is a law that all *Fs* are *Gs* if and only if (1) all *Fs* are *Gs*, and (2) that all *Fs* are *Gs* has a privileged epistemic status in our cognitive inquiry. But this purely epistemic account of lawlikeness fails to draw a robust line between laws and accidents.

The inference-ticket view: some empiricists argued that law-statements should not be seen as expressing propositions, and hence as being amenable to claims of truth and falsity. Rather, they must be seen as disguised rules of **inference**. We cannot validly move from the singular claim that '*a* is *F*' to the singular claim (perhaps, prediction) that '*a* is *G*', unless we use the sentence 'All *Fs* are *Gs*'. On the inference-ticket view, the function of law-statements is exactly this: they entitle us to make inferences such as the above. This view was accepted by **Schlick** and **Ramsey** partly on the grounds that nomological statements were meaningless, because unverifiable. But apart from the bankruptcy of the **verifiability** criterion of meaning, it is difficult to see how a statement of the form 'All *Fs* are *Gs*' can serve as a premise in a valid **deductive argument** without having a truth-value.

The web-of-laws view: a much more promising attempt to characterise lawlikeness is what may be called the web-of-laws view: the regularities that constitute the laws of nature are those that are expressed by the axioms and theorems of an ideal deductive system of our knowledge of the world, and, in particular, of a deductive system that strikes the *best* balance between **simplicity** and strength. Simplicity is required because it disallows extraneous elements from the system of laws. Strength is required because the deductive system should be as informative as possible about the laws that hold in the world. Whatever regularity is not part of this *best system* it is merely accidental: it fails to be a genuine law of nature. The gist of this approach, which has been advocated by **Mill**, and in the twentieth century by **Ramsey** in 1928 and **Lewis** in 1973, is that no regularity, taken in isolation, can be deemed a law of nature. The regularities that constitute laws of nature are determined in a kind of holistic fashion by being parts of a structure. Though the Mill–Ramsey–Lewis view has many attractions, it faces the charge that it cannot offer a fully *objective* account of laws of nature. But there is nothing in the web-of-laws approach that makes laws mind-dependent. The regularities that are laws are fully objective, and govern the world irrespective of our knowledge of them, and of our being able to identify them.

The necessitarian view: in the 1970s, David Armstrong (born 1926), Fred Dretske (born 1932) and Michael Tooley (born 1941) put forward the view that lawhood cannot be reduced to regularity. Lawhood, they claimed, is a certain contingent necessitating relation among properties (**universals**). Accordingly, it is a law that all *Fs* are *Gs* if and only if there is a relation of nomic necessitation $N(F, G)$ between the universals *F*-ness and *G*-ness such that all *Fs* are *Gs*. This approach aims to explain why there are regularities in the world: because there are necessitating relations among **properties**. It also explains the difference between nomic regularities and accidents by claiming that the accidental regularities are not even symptoms of the instantiation of laws. But the central concept of nomic necessitation is still not sufficiently clear. In particular, it is not clear how the necessitating relation between the property of *F*-ness and the property of *G*-ness makes it the case that *All Fs are Gs*. To say that there is a necessitating relation $N(F, G)$ is not yet to explain what this relation is.

The metaphysical contingency of laws: reacting to the Cartesian view that the laws of nature – and especially the fundamental ones – were metaphysically necessary and knowable a priori (since they were supposed to stem directly from the immutability of God), Humean-empiricists argued that laws of nature have to be contingent since they cannot possibly be known **a priori**; the actual laws do not hold in all possible worlds and there could be different laws holding in the actual world. This view that laws are contingent was adopted by the advocates of lawhood as a necessitating relation among universals. According to this view, the relation of nomic necessitation does *not* amount to logical (or metaphysical) necessity. There may be possible worlds in which $N(F, G)$ does not hold. Besides, nomic connections among universals are discoverable only a posteriori. No amount of a priori reasoning could establish that $N(F, G)$ holds.

The metaphysical necessity of laws: a growing rival thought has been that if laws did not hold with some kind of objective necessity, they could not be robust enough to support either causation or explanation. As a result of this, laws of nature are said to be metaphysically necessary. This amounts to a radical denial of the contingency of laws. The advocates of metaphysical necessity take the line that laws of nature flow from the essences of **properties**. In so far as properties have essences, and in so far as it is part of their essence to endow their bearers with a certain behaviour, it follows that the bearers of properties *must* obey certain laws, those that are issued by their properties. The thought that laws are metaphysically necessary gained support from the (neo-Aristotelian) claim that properties are active **powers**. On this view, properties are not freely recombinable: there cannot be worlds in which two properties are combined by a different law than the one that unites them in the actual world. Hence, it does not even make sense to say that properties are united by laws. Rather, properties – *qua* powers – *ground* the laws.

See **Descartes; Kant; Leibniz; Natural kinds**

Further reading: Armstrong (1983); Carroll (1994); Lange (2000); Mumford (2004); Psillos (2002)

Laws of thinghood: Intuitive laws that an entity should satisfy to be a **particular** thing (as opposed to a **universal**). A particular thing cannot be wholly present at two different places at the same time. And two or more particulars cannot occupy the same place at the same time. A universal violates both of these laws. A universal can be wholly present at two different places at the same time; and two (or more) universals can occupy the same place at the same time.

Further reading: Armstrong (1989)

Leibniz, Gottfried Wilhelm (1646–1716): German philosopher and mathematician, who invented the differential and integral calculus (independently of **Newton**). His best-known philosophical works are *Discourse on Metaphysics* (1686), *New Essays on Human Understanding* (1705) and *The Monadology* (1714). He drew a distinction between two kinds of **truth**: truths of reason, which are necessary because their negation implies contradiction, and truths of fact, which are contingent since their negation describes a possible state of affairs. He also drew a distinction between the phenomenal world and a deeper metaphysical reality behind it – the world of substances or *monads*. These are mind like, non-interacting, non-extended and simple substances on which the world of the phenomena (the world of matter in motion) is grounded. Leibniz thought that the Cartesian view that the essence of matter was extension was incorrect. Extension cannot account for the presence of activity in nature. In *Discourse on Metaphysics*, he argued that the essence of substance is activity. Though Leibniz favoured mechanical explanations of natural phenomena and denounced occult qualities as non-explanatory, he was not content with the prevailing mechanistic explanations of phenomena. He thought that the mechanical principles of nature need metaphysical grounding and that they should be supplemented by dynamical explanations in terms of forces and powers. Like **Descartes**, Leibniz thought that the fundamental **laws of nature** stemmed directly from God. Yet he drew a distinction between the most fundamental law of nature, namely, that nature is orderly and regular, and subordinate laws such as the laws of motion. According to Leibniz, the universal law of the general order is metaphysically necessary, since in whatever way God might have created the world it would have been orderly and regular. The basic Leibnizian laws of motion, such as the law of conservation of *vis viva*, were conservation laws. Hence, being invariant, they preserve the fundamental order of nature. The subordinate laws are metaphysically *contingent*, since they might differ in other possible worlds. Ultimately, all natural laws are explained by means of two central Leibnizian principles: *the principle of sufficient reason* and *the principle of fitness*. According to the first, for everything that happens, there must be a reason, sufficient to bring this about instead of anything else. According to the second, the actual world is the fittest or most perfect among all possible worlds that God could have created. Leibniz did admit teleological explanations alongside mechanical ones. In the end, all things have efficient *and* final causes. Leibniz's reconciliation is effected by means of a third principle, the *principle of pre-established harmony*. In all its generality, this principle states that when God created this world as the best among an infinity of possible worlds, he put everything in harmony (the monads and the phenomenal world, the mind and the body, the final and the efficient causes). In 1715–16, Leibniz was involved in a heated correspondence with Samuel Clarke (1675–1729), a leading Newtonian philosopher who represented Newton's views, which covered a number of philosophical issues from **space** and **time** to the nature of miracles. This was Leibniz's second controversy with Newton, the first concerning the invention of the calculus.

Further reading: Leibniz (1973)

Lewis, David (1941–2001): American philosopher, one of the most influential of the twentieth century. He did work in most areas of philosophy. He authored *Counterfactuals* (1973) and *On the Plurality of Words* (1986). He has been a defender of Humeanism – understood as denying the existence of **necessary connections** in nature. He defended **Humean supervenience** and claimed that natural laws are regularities. He was a modal realist, thinking that other possible worlds are no less real than the actual. He found in modal realism the resources to tackle a number of philosophical problems, including the nature of **properties**. Lewis advanced a theory of causal **explanation** and the counterfactual approach to **causation**. Perhaps, his most important contribution to the philosophy of science per se was his work on theoretical terms and his use of **Ramsey-sentence**s in specifying their meaning.

See **Counterfactual conditionals; Laws of nature**
Further reading: Lewis (1999); Nolan (2005)

Likelihood: Technical term for the **conditional probability** of the **evidence** given the hypothesis. If the hypothesis entails the evidence, the likelihood of the evidence given the hypothesis (i.e., prob(e/H)) is one.

Further reading: Sober (2002)

Likelihoodism: It uses the **likelihood** ratio (prob(e/H$_1$)/ prob(e/H$_2$)) to capture the strength by which the **evidence** supports one hypothesis over another, but it does not license judgements as to what the **probability** of a hypothesis in the light of the evidence is. Given two hypotheses H$_1$ and H$_2$, and evidence e, likelihoodism tells us that e supports H$_1$ more than H$_2$ if prob(e/H$_1$)>prob(e/H$_2$). The likelihood ratio prob(e/H$_1$)/prob(e/H$_2$) is said to capture the strength of the evidence. Compared to **Bayesianism**, likelihoodism is a modest philosophical view. It does not require the determination of prior probabilities. So it does not specify posterior probabilities and hence does not tell us what to believe or which hypothesis is probably true in the light of evidence.

See **Probability, posterior; Probability, prior**

Further reading: Hacking (1965); Sober (2002)

Literal interpretation: An interpretation (assignment of meaning to the terms and predicates) of a theory that takes it at face-value, that is an interpretation of a theory that does not re-interpet its claims as being about some domain other than that implied by a face-value reading of the theory. A literal interpretation of, say, the theory of electrons, takes the theory to be about *electrons* and their properties and refrains from re-interpreting the theory as being, say, about observable entities and their actual and possible behaviour. A literal interpretation is contrasted to figurative interpretation, which reads the theory as a metaphor. It is also contrasted to a reductive inter-pretation which takes the truth-conditions of the claims of the theory to be fully specified in a vocabulary other than the one used in the theory. **Scientific realism** takes scientific theories literally, while **reductive empiricism** has tried to offer empiricism-friendly reductive interpreta-tions of them.

See **Constructive empiricism; Fictionalism**

Further reading: Psillos (1999); van Fraassen (1980)

Locke, John (1632–1704): English philosopher, author of *An Essay Concerning Human Understanding* (1689). He adopted **empiricism** and **nominalism**. He thought that all ideas come from impressions and claimed that what-ever exists is **particular**. He claimed that **universals** are not real but inventions of the human mind. He adopted as fundamental the distinction between primary and sec-ondary qualities. The former (solidity, extension, figure, motion/rest and number) are real qualities and utterly in-separable from the body: they are the real substructure of the body. Secondary qualities, in contrast, are **powers** of the body to produce various sensations in our minds. They are produced by the operations of the invisible parti-cles of bodies on our senses. Locke also drew a distinction between real essences and nominal ones. The real essence of a thing is its underlying internal constitution, based as it is on its primary qualities. The nominal essence con-cerns the observable characteristics of things and amounts to an artificial constitution of a genus or a species. The nominal essence of gold, for instance, is a body that is yel-low, malleable, very soft and fusible. Its real essence is its microstructure. Being a nominalist, Locke thought that real essences are individuals, whereas nominal essences are mere **concepts** or ideas that define a species or a kind. Locke did not claim that real essences were unknowable but he was pessimistic about the prospects of **knowledge** about them. Though knowledge of nominal essences was possible, Locke thought that this kind of knowledge was trivial and uninteresting.

Further reading: Locke (1689)

Logical positivism: School of thought associated with the **Vienna Circle**. It came to be known as *logical positivism* because it put together the positivist demand that all synthetic **knowledge** should rest on experience and be the product of positive methods with the idea that philosophy is logical analysis, and in particular the logical analysis of the language and the basic **concepts** of science. Among its central doctrines was the rejection of the possibility of synthetic **a priori** knowledge; a conventionalist approach to logic and mathematics; the **verifiability** criterion of meaningfulness; the radical critique of metaphysics and the new, scientific, method of philosophising (based as it was on logic). The school was by no means monolithic and solid. Its members were involved in occasionally heated debates about all major philosophical issues, notably about whether knowledge needs foundations, whether scientific hypotheses are verifiable or simply confirmable, the nature of **truth,** the locus of **objectivity** etc. Though some basic tenets of **empiricism** were never questioned by the logical positivists, their thought, notably their claim that some framework principles (especially logical and mathematical ones) were necessary for experience, had a distinctively Kantian origin.

See **Analytic/synthetic distinction; Carnap; Neo-Kantianism; Neurath; Protocol sentences; Schlick; Unity of science; Verifiability**

Further reading: Ayer (1959); Friedman (1999); Giere and Richardson (1996)

Logicism see **Frege**

Lottery paradox: Imagine a fair lottery with *n* tickets and suppose, for simplicity, that each ticket is sold to different people. One of them will be the winner. The **probability** that an arbitrary ticket will *not* win is $1-1/n$. (If there are 1,000 tickets, the probability that an arbitrary ticket will not win is 0.999.) Suppose we reason thus: since the probability that my ticket will not win is almost one, my ticket (say, ticket no. 1) will not win. Suppose all *n* holders of tickets reason as above. Then, it follows that: ticket no. 1 will not win *and* ticket no. 2 will not win *and ... and* ticket no. 1,000 will not win. Hence, it follows that no ticket will win, though we know that one of them *must* win. This **paradox** is connected with the debate about **rules of acceptance**.

   Further reading: Kyburg (1974)

M

Mach, Ernst (1838–1916): Austrian physicist and philosopher of science. In 1895 he was appointed Professor of the Philosophy of Inductive Science in the University of Vienna. His philosophical views were mostly presented through his scientific treatises, most notably his *The Science of Mechanics* (1883). He rejected the Newtonian absolute **space** and **time** because they were unobservable. He argued that 'science is economy of thought' and that its aim is to classify appearances in a concise and systematic way. He thought it was not the business of science to posit unobservables which can *explain* the behaviour of phenomena. Mach rejected **atomism** on the basis of the claim that the positing of atoms is not a continuous extrapolation from the phenomena. He advocated phenomenological physics and claimed that all attempts to go beyond the sensory facts were metaphysical. Mach's anti-metaphysical and positivistic views exerted strong influence on many scientists and philosophers of science, including **Poincaré**, **Einstein**, the **Vienna Circle** and Bridgman.

See **Nominalism; Operationalism**

Further reading: Mach (1910)

McMullin, Ernan (born 1924): American philosopher of science, author of a number of influential articles on scientific realism, the history of philosophy of science, rationality and other areas and of *The Inference That Makes Science* (1992). He has defended **scientific realism** against **constructive empiricism** and has stressed the link between explanatory considerations and rational **belief**. McMullin has argued that **inference to the best explanation** (what he calls 'retroduction') is the inference that makes science. More recently, he has highlighted the role of values in scientific theorising.

See **Theoretical virtues**

Further reading: McMullin (1992)

Markov condition see **Causal graphs**

Materialism see **Nagel; Physicalism; Smart**

Maxwell, Grover (1918–1981): American philosopher of science who succeeded **Feigl** as the Director of the Minnesota Center for Philosophy of Science. He is famous for his defence of the reality of the theoretical entities posited by **scientific theories**. He argued that observability is a vague notion and that, in essence, *all* entities are observable under suitable circumstances. Maxwell also resuscitated **Russell's structuralism** and combined it with the **Ramsey-sentence** approach to scientific theories. He defended **structural realism** as a form of representative **realism**, which suggests that (1) scientific theories issue in existential commitments to **unobservable entities** and (2) all non-observational **knowledge** of unobservables is *structural knowledge*, that is, knowledge not of their first-order (or intrinsic) **properties**, but rather of their higher-order (or structural) properties.

Further reading: Maxwell (1962)

Maxwell, James Clerk (1831–1879): Scottish scientist, the founder of electromagnetism. Before him, it was assumed that electric and magnetic actions are propagated at-a-distance. Hence, the possibility that light might be of the same kind as electric and magnetic action was not taken seriously, since light was known to travel with finite velocity. Maxwell's fundamental discovery was that light is an electromagnetic wave propagated through the electromagnetic field according to electromagnetic laws. This discovery was based on the use of mechanical **models**, most notably the so-called *idle wheels* model. Though Maxwell freely used models and analogies, he was careful to point out that no **analogy** – no matter how suggestive and useful it might have been – was a real surrogate for a mature explanatory theory. This theory was first introduced in 'On the Dynamical Character of the Electromagnetic Field' in 1864 and was fully developed in *Treatise on Electricity and Magnetism*, in 1873, where Maxwell based his theory of the electromagnetic field on the general principles of dynamics and also derived the equations of the electromagnetic field.

Further reading: Maxwell (1890)

Meaning holism see **Holism, semantic**

Mechanical philosophy: View of the world and of science characteristic of much of the seventeenth century and beyond, according to which all natural phenomena are explainable mechanically in terms of matter in motion. It took efficient **causation** (which it understood as pushings and pullings) as the only form of causal interaction and either excised all final causation from nature or placed it firmly in the hands of God. Though the broad contours of mechanical philosophy were not under much dispute, the specific principles it was supposed to endorse were heavily debated. Some mechanical philosophers (notably Pierre Gassendi, 1592–1655) subscribed to **atomism,** while others (notably **Descartes**) took the universe to be a plenum, with matter being infinitely divisible.

See **Boyle; Leibniz; Mechanism**

Further reading: Losee (2001); Wilson (1999)

Mechanism: The idea that nature forms a mechanism was part of the **mechanical philosophy**. A mechanism was taken to be any arrangement of matter in motion, subject to the laws of mechanics. More specifically, it was thought that all macroscopic phenomena were the product of the interactions (ultimately, pushings and pullings) of microscopic corpuscles. The latter were fully characterised by their primary qualities. A mechanical **explanation** was taken to lay bare the mechanism that produces a certain effect. With the advancement of science, the content of mechanism was broadened. After **Newton**, a new category, *force*, was introduced alongside the two traditional mechanical categories, *matter* and *motion*. Mechanical explanation was taken to consist in the subsumption of a phenomenon under Newton's laws. In the nineteenth century, when the issue of the possibility of mechanical explanation of electromagnetic phenomena was discussed, **Poincaré** suggested that a necessary and sufficient condition for a mechanical explanation of a set of phenomena is that there are suitable potential and kinetic energy functions such that they satisfy the principle of conservation of energy. Given that such functions can be specified, there will be a configuration of matter in motion (actually, Poincaré showed that there will be indefinitely many such configurations) that can underpin a set of phenomena. But then, Poincaré thought, it is not the search for mechanisms that is important, but rather the search for *unity* of the phenomena under laws of conservation. In the twentieth century, the search for mechanisms and mechanical explanations was a weapon against **vitalism**. But in the philosophy of explanation, the search for mechanisms gave way to the claim, captured by the **deductive-nomological model of explanation**, that subsumption under laws is enough for explanation. Recently, there has been a resurgence of interest in mechanisms, partly because of developments in the sciences and partly because of the failures of the standard models of explanations. A view that gains support is that **causation** is best understood in terms of mechanisms that connect cause and effect. There have been two broad ways to understand mechanisms. The first (defended by **Salmon**) is to take mechanisms to be processes, and in particular **causal processes**. The second is to take mechanisms to be complex objects (systems), that is, stable arrangements of entities that perform a certain **function** and are understood by reference to the **properties** and interactions of their component parts.

See **Descartes; explanation, unification model of; Leibniz**

Further reading: Glennan (2002); Machamer, Darden and Craver (2000); Salmon (1984)

Mellor, David Hugh (born 1938): British metaphysician, author of *Matters of Metaphysics* (1991). He has further articulated and defended the views of **Ramsey**. He has taken **chance** as a key conceptual category in understanding the world, has defended the view that all properties are **dispositions** and has highlighted the interconnectedness of laws and properties. According to Mellor, chances are tendencies of actual particulars, and they exist as real properties with a definite causal role. **Laws of nature** embody relations among properties: they express the chance that the instantiation of a property will lead to the instantiation of another. **Properties** are identified a posteriori by looking at the **Ramsey-sentences** of scientific theories.

Further reading: Mellor (1991, 1995)

Methodological naturalism: The view that methodology is an empirical discipline and that, as such, it is part and parcel of natural science. It suggests the following. (1) Normative claims are instrumental: methodological rules link up aims with methods which will bring them about, and recommend what action is more likely to achieve one's favoured aim. (2) The soundness of methodological rules depends on whether they lead to successful action, and their **justification** is a function of their effectiveness in bringing about their aims. A sound methodological rule recommends the best strategy for reaching a certain desired aim.

See **Axiology; Giere; Naturalism**

Further reading: Giere (1988); Laudan (1996)

Mill, John Stuart (1806–1873): English philosopher, author of *A System of Logic Ratiocinative and Inductive* (1843). He was an advocate of a radical **empiricism** and **inductivism**, according to which all **knowledge** (even in mathematics and geometry) is grounded in experience. He rejected the possibility of **a priori** knowledge and claimed that all knowledge was ultimately inductive. Induction,

Mill thought, is conceptually prior to deduction since valid **deductive arguments** rely on universal generalisations that can only be established inductively. He thought that the laws of logic were empirical laws and argued that, being ultimately the most general laws of nature, they are grounded in experience. By tying mathematical **truth** to experience, Mill thought that the content of mathematical statements was the empirical world. He thought that though **induction** can only be justified empirically, it cannot be really doubted since, even after we have reflected on the issue of its justification, we cannot help relying on it. But he also claimed that induction is supported by its empirical successes; and, in particular, by a second-order induction that leads to the conclusion that all phenomena fall under regularities. This, he thought, was the law of universal causation. He was a defender of the regularity view of **causation**, with the sophisticated addition that the cause of an effect should be taken to be the whole conjunction of the conditions that are sufficient and necessary for the effect. For Mill, regular association is not, on its own, enough for causation. A regular association of events is causal only if it is 'unconditional', that is, only if its occurrence does not depend on the presence of further factors which are such that, given their presence, the effect would occur even if its putative cause was not present. Mill tried to delineate the **scientific method** in such a way that it can lead to causal knowledge. He put forward the *Method of Agreement* and the *Method of Difference*. According to the first, the cause is the common factor in a number of otherwise different cases in which the effect occurs. According to the second, the cause is the factor that is different in two cases, which are similar except that in the one the effect occurs, while in the other it doesn't. Mill became involved in a debate with **Whewell** concerning the role of **novel predictions**. Unlike

Whewell, Mill thought that no predictions could *prove* the truth of a theory. He added that novel predictions carry no extra weight over predictions of known facts. Mill should also be credited with the first attempt to articulate the **deductive-nomological model of explanation**, which became prominent in the twentieth century. The explanatory pattern that Mill identified is deductive, since the explananda (be they individual events or regularities) must be deduced from the explanans. And it is nomological, since the explanans must include reference to laws of nature. He also took **unification** to be the hallmark of explanation and of laws. Unification is explanatory because it minimises the number of laws that should be taken as ultimately mysterious, that is, as inexplicable. This very process of unification, Mill thought, brings us nearer to solving the problem of what the **laws of nature** are. They are the fewest general propositions from which all regularities that exist in nature can be deduced.

Further reading: Mill (1911)

Mill's methods see **Eliminative induction; Mill; Scientific method**

Mind-independence see **Idealism; Objectivity; Realism and anti-realism; Scientific realism**

Models: Term of art used in understanding how theories represent the world. Though according to a popular view, the **semantic view of theories**, theories are families of models, there is little agreement as to what models are, how they are related to theories and how they represent whatever they are supposed to represent. In the first half of the twentieth century, where the **syntactic view of theories** ruled, models were taken to be conceptual devices which cast the theory in familiar terms, thereby facilitating its understanding. For instance, the billiard-ball model of gases, which conceived of molecules as perfectly elastic spheres, was supposed to offer a familiar picture of the kinetic theory of gases. Besides, models were taken to help ground the interpretation of the theory in experience. This view was challenged by what may be called the analogical approach to models, championed mainly by **Achinstein** and **Hesse** in the 1960s, who focused their attention on models of physical systems. A theoretical model of a target physical system X is taken to be a set of theoretical assumptions (normally of a complex mathematical form) which provide a starting point for the investigation of the behaviour of the target system X, where the choice of assumptions is guided by substantive similarities (analogies) between the target system X and some known physical system Y. **Suppes** initiated a new approach to models by taking them in the logician's sense: a model is a **structure** that makes a theory true. Suppes insisted that the concept MODEL has the same meaning in mathematics and empirical science and argued that a theory should be construed as a set of abstract structures, namely, a set of models that render the theory true. He favoured, as he put it, an extrinsic characterisation of theory, whereby to present a theory is to define the intended class of models of the theory. Suppes shifted attention from models of physical systems (i.e., analogical or iconic models) to models of theories, that is, mathematical models. According to **Cartwright**, models are devices employed whenever a mathematical theory is applied to reality. This view has recently been developed into the models-as-mediators programme, according to which models are autonomous agents that mediate between theory and world.

See **Analogy**

Further reading: Morgan and Morrison (1999)

Musgrave, Alan (born 1940): English-born New Zealander philosopher of science, editor (together with **Lakatos**) of *Criticism and the Growth of Knowledge* (1970) and author of *Essays on Realism and Rationalism* (1999). He has defended **scientific realism**, which he took it to be, by and large, an axiological thesis: science aims for true theories. Being a Popperian, he claimed that the **truth** of **scientific theories** can never be established (nor can it be made probable), but that it always remains conjectural. He subscribes to **deductivism** and denies that there are cogent non-deductive arguments. More recently, he has argued that **critical rationalism** should be accompanied with a sort of **voluntarism**. The idea is that one can be reasonable in believing some proposition p (e.g., that a theory is true) even if the evidence that there is for it does *not* raise its probability of being true.

Further reading: Musgrave (1999)

Nagel, Ernest (1901–1985): American philosopher, author of *The Structure of Science: Problems in the Logic of Scientific Explanation* (1961) and of *Teleology Revisited* (1982). Early in his career, he was an advocate of **naturalism**, arguing that the **scientific method** is the most reliable way to achieve knowledge, and of non-reductive materialism, arguing that there are logically contingent causal connections between mental states and physical ones. Later on he took a positivist turn, arguing, for instance, that **realism** and **instrumentalism** are merely *different languages* about theories and the choice between them is only a choice of the preferred mode of speech. He elaborated the **deductive-nomological model of explanation** and, based on it, he developed a theory of **reduction** that became the standard account for many decades. He also tried to reconcile teleological explanations with causal ones.

See **Functional explanation**

Further reading: Nagel (1960, 1977)

Natural kinds: Categories of things (or stuff) that are supposed to share something in common in virtue of which they form a kind. For instance, electrons form a natural kind, and so does water and gold and cats. According to a strong view, the members of a kind share in common the same essence (i.e., the same essential properties). This shared essence is taken to be an objective characteristic of the members of the kind (a Lockean real essence) which determines the salient **properties** of the members of the kind. It is supposed to ground the modal features of kind-membership: that the members of the kind have, necessarily, some properties. On this strong view, which may be called **essentialism** about kinds, kinds are discrete: no entity can belong to two distinct fundamental kinds. These discrete boundaries of kinds are supposed to form the joints of nature (as Plato put it) and the aim of science is taken to carve nature at its joints – that is, to uncover the objective natural kind structure of the world. According to a weaker view, there are natural kinds in nature but kind-membership is not a matter of sharing essential properties but a matter of objective similarities and differences among the members of the kind. This view allows that kinds might differ in degree and not absolutely from one another. A more radical view takes it that natural kinds are conventional constructions that have to do with our own classificatory schemata. The general criticism of essentialism (especially about biological kinds – species – which defy an essentialist characterisation) led the essentialist conception of natural kinds into disrepute. But **Kripke's** and **Putnam's causal theory of reference** and the progressive rehabilitation of essentialism has led to the revival of essentialist conceptions of natural kinds. Essentialism has now become dispositional, taking the line that the essential kind-constitutive properties are causal **powers** of things. An important view developed by **Boyd,** which defies essentialism without abandoning **realism** about natural kinds, is that kinds are homeostatic property clusters. No matter what exactly one thinks about natural kinds, they have played a key role in many philosophical issues, such as the problem of **induction,** the **laws of nature, reduction, confirmation** and **explanation.**

See **Essentialism, dispositional; Grue; Incommensurability**

Further reading: Bird (1998); Wilkerson (1995)

Natural ontological attitude: Stance towards science advanced by **Fine.** It rejects philosophical theories about science, be they realist or anti-realist, as unnatural attachments to science on the basis that they try to authenticate science. While **realism** aspires for an outside authentication of science, by taking science to be about the world, anti-realism aims for an inward authentication of science, by taking science to be about humans and our relations with the observable world. The natural ontological attitude takes science at face value and seriously without trying to interpret it; nor to offer a metaphysical or epistemological foundation to it. More specifically, the natural ontological attitude claims that the concept of **truth** employed in science needs no philosophical interpretation.

Further reading: Fine (1986)

Naturalism: Cluster of views that puts natural science, its method and its findings at the centre of any attempt to understand the world and our relationship with it. It puts the mind firmly within the world and denies that there are special mental faculties by means of which **knowledge** of the world is possible. It is associated with **Hume** and **Mill**, but it was eclipsed by **Kant**'s transcendentalism. Its re-appearance in the twentieth century rested on two pillars: the denial of the synthetic **a priori** and the defence of psychologism. In denying the very possibility of a priori knowledge, naturalism denies any special cognitive or methodological status to philosophy. This naturalist view was the central theme of **Quine**'s influential paper 'Epistemology Naturalised' (1969). In this, Quine argued that, once the search for secure foundations of knowledge is shown to be futile, philosophy loses its presumed status as the privileged framework (equipped with a privileged source of knowledge: a priori reflection and logical analysis) aiming to validate science. Philosophy becomes continuous with the sciences in the sense that there is no privileged philosophical method, distinct from the **scientific method**, and that the findings of empirical sciences are central to understanding philosophical issues and disputes. Quine went as far as to suggest a replacement thesis: that epistemology, as traditionally understood, should give way to psychology. Quine made capital on the vivid metaphor of **Neurath's boat**. In favouring psychologism, naturalism rejected the apsychologistic character of traditional epistemology and philosophy of science, which aimed at a logical analysis of key concepts independently of the psychological and social processes by which they are implemented. **Methodological naturalism** was an attempt to show how scientific methodology can be justified in a broadly empirical way. Though some naturalists restricted their naturalism to methodology and epistemology, others took naturalism to be a metaphysical doctrine. It restricts what there is to whatever is in **space** and **time** and makes causal contributions to the workings of the world. As such, it exerts pressure on whatever is prima facie non-natural (namely, the mental, the moral, the mathematical, the evaluative, the justificatory etc.) to *earn its right* to be included in the natural world. It goes without saying that naturalism excludes supernaturalism. Naturalism has been challenged on the grounds that: (1) it is circular; (2) it cannot recover the normative judgements which traditional epistemology was supposed to deliver; and (3) it falls prey to **relativism**.

See **Physicalism**
Further reading: Papineau (1993); Quine (1969)

Necessity: see **Analytic/synthetic distinction; A priori/a posteriori; Essentialism; Hume; Kripke; Laws of nature; Necessary connection**

Necessary connection: What **Hume** searched for in causal sequences, but could not perceive: a tie that links cause and effect in virtue of which the cause brings about the effect; or makes it inevitable that the effect will happen; or necessitates the effect. Traditional accounts of **causation** had assumed that there are necessary connections in nature. Hume did admit that the idea of necessary connection is part of the ordinary concept of cause but tried to explain its origin by showing how it is projected onto nature by the human mind.

See **Causation; Induction, the problem of**
Further reading: Hume (1739); Psillos (2002)

Neo-Kantianism: Philosophical current aiming to adapt Kant's thought to the developments in the sciences during the nineteenth century. It has been divided into two schools: the school of Marburg and the Southwest German School of Baden. The main members of the Marburg school were Hermann Cohen (1842–1918), Paul Natorp (1854–1924) and Ernst Cassirer (1874–1945). It is characterised by its attention to logic and the natural sciences. It took mathematics and the natural sciences as the models of knowledge and denied the key Kantian thought that **knowledge** has a double source: **concepts** *and* intuition. Cassirer was the most eminent neo-Kantian with strong influence on **logical positivism**. He criticised **empiricism** on the basis that knowledge requires the existence of structures (**space, time,** relations) that put order to experience and argued that logic and mathematics provide these structures. In *Substance and Function* (1910), he argued that, though the phenomena could be identified, organised and structured *only* if they were embedded in mathematical structures, these structures were not fixed for all time and immutable. He thought that mathematical structures, though synthetic **a priori** – since they are required for objective experience – are revisable, yet convergent, since newer structures accommodate within themselves old ones. The main members of the Southwest school were Wilhelm Windelband (1848–1915) and Heinrich Rickert (1863–1942). It is characterised by its focus on values and their role in knowledge. It turned its attention to history and the human sciences and aimed to unveil their peculiarities vis-à-vis the natural sciences. Windelband introduced a distinction between the idiographic method, which characterises the human sciences, and is focused on singular events and their connections and the nomothetic method, which characterises the natural sciences and aims at general judgements and lawful connections. The nomothetic approach tends to abstraction and is value free, whereas the idiographic approach tends to the concrete and is value laden.

Further reading: Cassirer (1910)

Neurath, Otto (1882–1945): Austrian philosopher, sociologist and political activist, one of the most radical members of the **Vienna Circle**. He is mostly known for his critique of **foundationalism** – especially in the **protocol sentences** debate. He defended a version of **coherentism** about **justification** and argued that no statement is immune to revision. He claimed that no statement could be compared to the facts – that statements can only be compared to other statements – a view that many took to imply an account of **truth** as coherence. He also defended **physicalism** – as a doctrine about the unity of the language of science. His metaphor (**Neurath's boat**) became one of the defining intuitions of **Quine**'s naturalised epistemology. After the *Anschluss* Neurath escaped first to Holland, and then to England, where he worked for a public housing authority. He was one of the founders of the *International Encyclopedia of Unified Science* and the associated movement for the **unity of science,** where the ideas of **logical positivism** came in contact with American **pragmatism**.

Further reading: Neurath (1983)

Neurath's boat: Vivid metaphor introduced by **Neurath** to boast **coherentism**. It became famous in the writings of **Quine** as the constitutive metaphor of **naturalism**. Neurath claimed that in our trying to investigate how well we are doing in our cognitive give-and-take with the world we are like sailors 'who have to rebuild their ship on the open sea, without ever being able to dismantle it in drydock and reconstruct it from the best components'. Quine used this metaphor to argue against first philosophy, namely, the view that philosophy has a foundational role vis-à-vis the sciences, aiming to validate them and secure their claim to knowledge. Philosophy has no special status; any parts of our conceptual scheme (the findings of science in particular) can be relied upon when revisions elsewhere in our conceptual scheme are necessary. Since there is no dry dock in which we can place our conceptual scheme as a whole and examine it bit by bit, we are engaged in a process of mutual adjustment of its pieces while keeping it afloat.

Further reading: Quine (1960)

Newton, Isaac (1642–1727): One of the most famous scientists of all time, author of *Philosophiae Naturalis Principia Mathematica* (*Mathematical Principles of Natural Philosophy*, 1687). Apart from his well-known scientific achievements, Newton had considerable impact on methodological matters. His famous dictum *Hypotheses non fingo* ('I do not feign hypotheses') was supposed to act as a constraint on what can be known: it rules out all those metaphysical, speculative and non-mathematical hypotheses that aim to explain, or to provide the ultimate ground of, the phenomena. Newton took **Descartes** to be the chief advocate of hypotheses of the sort he was keen to deny. His official suggestion for the method of science was that it is deduction from the phenomena. Newton's approach was fundamentally mathematical-quantitative. He did not subscribe to the idea that **knowledge** begins with an experimental natural history of the sort suggested by **Bacon**. Yet, the basic laws of motion do stem from experience. The empirically given phenomena that Newton started with were laws (e.g., Kepler's laws). Then, by means of mathematical reasoning and the basic axioms or laws of motion, further conclusions could be drawn, for example, that the inverse square law of gravity applies to all planets. Newton's methodological views were the subject of great debate among his contemporaries and his successors. His ban on hypotheses was criticised as being inconsistent with his own scientific theory.

See **Kant; Laplace; Locke; Whewell**
Further reading: Cohen (1985)

Nicod, Jean (1889–1924): French philosopher and mathematician, author of the essay 'The Logical Problem of Induction', which was published in French in 1923 and was translated into English in 1930. In this, he claimed that law-like generalisations are established as probable, if at all, by being confirmed by their favourable (positive) instances and are refuted by being invalidated by their unfavourable (negative) instances, and argued that induction by enumeration, that is, **confirmation** by repetition, is the fundamental form of **induction**.

See **Confirmation, Hempel's theory of; Paradox of the ravens**
Further reading: Nicod (1969)

Nominal vs real essence see **Locke**

Nominalism: The view that only particulars exist. Nominalists have argued that general terms and predicates are merely names for classifying **particulars** in terms of their similarities and differences. Realists, on the other hand, have claimed that **universals** are real entities referred to by general names and predicates, and argued that they are necessary for grounding the similarities and differences among particulars. Against this, nominalists have tried to accommodate some conception of **properties** (based on the idea that properties are, ultimately, classes of particulars) without admitting that they are universals. There have been two senses of nominalism: against the reality of universals and against the existence of **abstract entities**.

Though the two senses are distinct, the stronger version of nominalism combines them in affirming that everything that exists is particular *and* concrete, where 'particular' rules out universals and 'concrete' rules out abstract objects. Historically, nominalism has been associated with William of **Ockham** and **Locke**. It has been a bedfellow of **empiricism,** since the aversion to universals and abstract entities stems largely from an aversion to metaphysics and a commitment to the view that all **knowledge** should be grounded in sensory experience. Nominalism comes in several varieties.

Extreme nominalism: there are no properties (universals). Predicates apply to particulars, but they are just words which group together certain particulars.

Class nominalism: properties are classes of particulars and there is no further issue of why a certain particular belongs to a certain class. For instance, the property of redness is just the class of red things – the class to which all and only red things belong. Classes are particulars, since they are not repeatable: each class is defined by its members. An objection to class nominalism is that two predicates might have the same extension (i.e., they may apply to the same class of things) but capture different properties (e.g., 'is a renate' and 'is a cordate' apply to the same class of animals but designate different properties, namely, having kidneys and having a heart). Class nominalists argue that the extension of a predicate should be identified with actual *and* possible particulars. All renates are cordates in the actual world, and yet there are possible worlds in which particulars with hearts do not have kidneys; hence, the two predicates ('is renate' and 'is cordate') have different extension and form different classes.

Natural class nominalism: properties are *natural* classes of particulars. This is an attempt to meet the difficulties faced by class nominalism. The idea is that not all classes correspond to properties – only natural classes do. In some versions of it, the notion of 'natural class' is taken as primitive. Some argue that the very idea of a natural class is a precondition for our thinking about the world, since without it we cannot distinguish between the classes to which a certain particular belongs and the properties it possesses.

Resemblance nominalism: properties are classes of resembling particulars. This is an attempt to explain why some classes of particulars are natural while others are not. The idea is that there are similarities and differences among particulars in virtue of which they belong to classes. In some versions of resemblance nominalism, these similarities are objective features of the world – hence the natural classes are determined by objective worldly features (even though resemblance is always a matter of degree). In other versions of resemblance nominalism, the natural classes are the joint product of humans and nature. **Russell**'s argument against resemblance nominalism is that it has to posit at least one universal, namely, resemblance.

Nominalism and causation: A chief argument against nominalism, discussed in the Middle Ages and having resurfaced recently, is that properties – *qua* universals – are needed for understanding **causation** and laws. It is argued that things cause anything to happen in virtue of their properties and that only if properties are seen as universals can it be understood how they can enter into causal relation. For instance, it is claimed that it does not matter to the causal powers of a particular that it belongs to a certain class. Anti-nominalists also argue that **laws of nature** are best understood as relations among universals.

See **Natural kinds; Properties; Tropes**

Further reading: Armstrong (1989); Quine (1953); Quinton (1973)

No-miracles argument: Major and controversial argument in favour of **scientific realism**, also known as the ultimate argument for realism. It is based on **Putnam**'s claim that realism 'is the only philosophy of science that does not make the success of science a miracle'. Variants of it can be found in the writings of **Duhem** and **Poincaré** and, more recently, **Smart** and **Grover Maxwell**. **Boyd** and the author of this book have developed it into an argument for realism based on **inference to the best explanation**. No matter how exactly the argument is formulated, its thrust is that the success of scientific theories, and especially their ability to issue in **novel predictions**, lends credence to the following two theses: (1) that scientific theories should be interpreted realistically; and (2) that, so interpreted, these theories are approximately true. On a realist understanding of theories, novel predictions and genuine empirical success is to be expected. Critics of the no-miracles argument claim that it begs the question against non-realists since it relies on an inference to the best explanation, a mode of reasoning whose credentials are doubtful. They have also claimed that the **pessimistic induction** discredits the no-miracles argument.

Further reading: Boyd (1981); Psillos (1999); Putnam (1978); Smart (1963)

Non-Euclidean geometries: Alternatives to **Euclidean geometry** developed as rigorous geometrical systems in the nineteenth century. They deny Euclid's fifth postulate. Nikolai Ivanovich Lobachevsky (1792–1856) and János Bolyai (1802–1860) developed a geometry which assumed that an infinite number of lines parallel to a given line could be drawn from a point outside it, and this (hyperbolic) geometry was proved to be consistent. Bernhard Riemann (1826–1866) developed a consistent (spherical) geometry which assumed that no lines parallel to a given line could be drawn from a point outside it. These geometries were originally admitted as interesting mathematical systems. The Kantian thought that the geometry of physical space had to be Euclidean was taken as unassailable. Yet, **Einstein**'s General Theory of Relativity suggested that far from being flat, as Euclidean geometry required, **space** – that is *physical* space – is curved; actually, a space with variable curvatutre, the latter depending on the distribution of mass in the universe. All three geometries (Euclidean, Lobachevkyan and Riemannian) posited spaces of constant curvature: zero, negative and positive respectively. They all relied on the Helmholtz–Lie axiom of free mobility, which, in effect, assumes that space is homogeneous. According to Einstein's General Theory objects in **spacetime** move along geodesics whose curvature is variable.

Further reading: Torretti (1978)

Novel prediction: Typically, the prediction of a phenomenon whose existence is ascertained *after* a theory predicts its existence. On this temporal understanding of novelty, a novel prediction is always a prediction of a hitherto *unknown* phenomenon. Drawing from the fact that theories get support from explaining already known phenomena, many philosophers (most notably Worrall, **Zahar** and Jarrett Leplin) have stressed another sense of novelty, namely, use novelty. A prediction of an already known phenomenon by a theory *T* is use-novel *relative* to theory *T*, if no information about this phenomenon was employed (or required) during the construction of the theory that predicted it. There has been considerable debate about how exactly this last requirement should be understood. **Scientific realism** is typically associated with the claim that the best **explanation** of the ability of some scientific theories to yield novel predictions is that these theories are approximately true.

See **Ad hocness/ad hoc hypotheses; Prediction vs accommodation**

Further reading: Leplin (1997); Maher (1993)

Numbers see **Abstraction principles; Fictionalism, mathematical; Frege; Platonism, mathematical**

Objectivity: Opposed to subjectivity, it stands for whatever is independent of particular points of view, perspectives, subjective states and preferences. There are two distinct senses of objectivity, depending on how exactly the demand of independence is understood. The *first* is inter-subjectivity, understood as the 'common factor' point of view: the point of view common to all subjects. Thus understood, objectivity amounts to inter-subjective agreement. The *second* sense is radical objectivity, whatever is totally subject-independent. In particular, objectivity in the second sense is understood as mind-independence or knowledge-independence. When, for instance, it is said that certain entities have objective existence, it is meant that they exist independently of being perceived, or known etc. The concept of objectivity acquires more concrete content when it is applied to more specific cases, as, for instance, the objectivity of belief, the objectivity of scientific method etc. In such cases, objectivity is intimately connected with **truth** and **rationality**. The objectivity of **belief**, for instance, is taken to be a function of the methods (or processes) followed for the acquisition of the belief, where the methods should be such that they tend to yield truths. Or, the objectivity of **scientific method** has been taken to be the outcome of the fact that this method has a rational **justification**. An increasingly popular view connects objectivity with *invariance*: objective is whatever remains invariant under transformations, or under a change of perspective or point of view. A popular way in which this view has been developed has been in terms of structural invariance: the **structure** (or form) is objective, while the content (or matter) is subjective.

See **Carnap; Devitt; Feminist philosophy of science; Feminist standpoint; Popper; Structural realism**

Further reading: Nozick (2001); Wright (1992)

Observation, theory-ladenness of: The view that all observation is dependent on theories. It goes back to **Duhem** and his claim that observation in science is not just the act of reporting a phenomenon; it is the *interpretation* of a phenomenon in the light of some theory and other background beliefs. For Duhem, the theoretical interpretation that always infiltrates observation embeds (a description of) the observed phenomenon into the abstract, ideal and symbolic language of theory. This implies that different theories will confer different interpretations on some phenomenon. Hence, strictly speaking, the observed phenomenon is *not* the same if it is informed by different theories. Duhem's suggestion was that this situation is not problematic in so far as there are some commonly accepted background beliefs that advocates of competing theories can appeal to in the interpretation of observations. The interest in the theory-ladenness of observation resurfaced in the 1960s, this time drawing on a mass of empirical evidence coming from psychology to the effect that perceptual experience is theoretically interpreted. In the famous duck-rabbit case, for instance, one does not merely observe a shape composed of certain curved lines. One sees a rabbit *or* a duck. There is no purely perceptual experience, even though its theoretical interpretation is, largely, unconscious. **Hanson, Kuhn** and **Feyerabend** pushed the theory-ladenness-of-observation thesis to its extremes, by arguing that each theory (or paradigm) creates its own experiences; it determines the meaning of *all* terms that occur in it and there is *no* neutral language which can be used to assess different theories (or **paradigms**). This gave rise to issues concerning **incommensurability**.

See **Fodor; Holism, semantic; Terms; observational and theoretical**

Further reading: Arabatzis (2006); Duhem (1906); Hanson (1958); Kuhn (1962)

Observational terms see **Terms, observational and theoretical**

Occasionalism: The view that the only real cause of everything is God and that all causal talk that refers to worldly substances is a sham. Nicholas Malebranche (1638–1715) drew a distinction between real causes and natural causes (or occasions). A true cause is one such that the mind perceives a necessary connection between it and its effect. Since, he thought, the mind perceives the necessary connection only in God's will to bring about an effect, it is only God who is the true cause of anything. Natural causes are merely the occasions on which God causes something to happen. For Malebranche, since causation involves a **necessary connection** between cause and effect, and since no such necessary connection is perceived in cases of alleged worldly causation, there is no worldly causation: in the world there are only regular sequences of events, which strictly speaking are not causal. There is a sense in which **Hume**'s views on **causation** can be described as occasionalism minus God.

Further reading: Malebranche (1674–5)

Ockham, William of (c. 1285–1347): English medieval philosopher, mostly known for his **nominalism** and his enunciation of what came to be known as **Ockham's razor**. Being a nominalist, he denied the existence of **universals** and claimed that general terms or predicates refer to concepts that apply to many **particulars**. Ockham argued that there is no **a priori** necessity in nature's workings: God could have made things other than they are. Hence, all existing things are contingent. He denied that there are **necessary connections** between distinct existences and hence argued that there cannot be **justification** for inferring one distinct existence from another. Accordingly, all **knowledge** of things should come from experience. He claimed that there could never be certain causal knowledge based on experience, since God might have intervened to produce the effect directly, thereby dispensing with the secondary (material) cause. His central disagreement with **Aristotle** was about the content of first principles. Since he thought there was nothing in the world that corresponded to general concepts (like a universal), he claimed that the first principles are, in the first instance, about mental contents. They are about concrete individuals only *indirectly* and in so far as the general terms and concepts can be predicated of concrete things.

Further reading: Ockham (1990)

Ockham's razor: Methodological principle connected to the virtue of **simplicity** or parsimony: entities must not be multiplied without necessity (*Entia non sunt multiplicanda sine necessitate*). Though attributed to William of **Ockham**, this principle of parsimony was well known in his time. Robert Grosseteste (c. 1168–1253) had put it forward as the *lex parsimoniae*, or law of parsimony: 'nature operates in the shortest way possible'. This principle was not meant to offer a metaphysical insight into what there is. As Ockham himself observed, God could have made the world very complex. It was, however, accepted as a sound methodological principle, or, more strongly, as a theoretical principle: there always should be some sufficient reason for positing entities. In the hands of the emergent radical nominalists, this principle was supposed to warn against the positing of **unobservable entities** and, in particular, **universals** and **abstract entities**.

See **Nominalism**

Further reading: Sober (1990)

Old evidence, problem of: Difficulty for the Bayesian theory of **confirmation**, first identified by **Glymour**. Suppose that a piece of evidence e is already known (it is, that is, an *old* piece of evidence relative to the hypothesis H under test). Its probability is equal to unity ($\text{prob}(e) = 1$). Given **Bayes's theorem**, this piece of **evidence** does not affect at all the posterior probability ($\text{prob}(H/e)$) of the hypothesis given the evidence: the posterior probability is equal to the prior probability, that is, $\text{prob}(H/e) = \text{prob}(H)$. This, it has been argued, is clearly wrong, since scientists typically use known evidence to support their theories. Therefore, there must be something wrong with Bayesian confirmation. Bayesians have replied by adopting an account of the relation between theory and old evidence based on **counterfactual conditionals**. They argue as follows. Suppose that B is the relevant background knowledge and e is an old (known) piece of evidence – that is, e is actually part of B. In considering what kind of support e confers on a hypothesis H, we subtract *counterfactually* the known evidence e from the background knowledge B. We therefore presume that e is not known and ask: what would be the probability of e given $B - e$? This will be less than one; hence, the evidence e can affect (i.e., raise or lower) the posterior **probability** of the hypothesis.

See **Confirmation, Bayesian theory of**
Further reading: Earman (1992); Glymour (1980)

Operationalism: The doctrine, defended by P. W. Bridgman (1882–1961), that theoretical **concepts** should be defined by operational **definitions**. A consequence of this doctrine is that given that a certain physical magnitude (e.g., temperature) can be detected by means of several experimental procedures (e.g., by an air thermometer, or by an alcohol thermometer, or by a mercury thermometer), we end up with a multiplicity of concepts, each being defined by virtue of some specific experimental procedure. Bridgman chose to live with this oddity: there is not just one magnitude of, say, *temperature* and merely different ways to measure it, or to apply it to experiential situations. On his view, there is a multiplicity of different magnitudes which we *wrongly* characterise by a single concept: *temperature*.

See **Definition, operational; Einstein; Hempel**
Further reading: Bridgman (1927); Hempel (1965)

Overdetermination, causal: It occurs when there are two factors each of which is sufficient to bring about the effect, but none of them is necessary, since, even if the one was not present, the other factor would ensure the occurrence of the effect. For instance, two rocks are simultaneously thrown at a bottle and they shatter it. Cases such as this present a problem to the counterfactual theory of **causation**. For, though both throwings of the rocks caused the shattering, the effect is not counterfactually dependent on either of them, since if one rock had missed the bottle the other would still have shattered it. So there is causation without the cause being counterfactually dependent on the effect.

Further reading: Lewis (1973a); Mackie (1974)

Paradigm: The dominant characteristics of a paradigm, as this was conceived by **Kuhn**, are: (1) it stands for the whole network of theories, beliefs, values, methods, objectives, professional and educational structure of a scientific community; and (2) it stands for a set of explicit guides to action (what Kuhn sometimes calls 'rules'). The later Kuhn replaced the single concept of a paradigm by two others: disciplinary matrix and exemplars. The *disciplinary matrix* includes: (1) the symbolic generalisations that a scientific community accepts as characterising the **laws of nature** or the fundamental equations of theories (2) the set of heuristic devices and analogies (**models**) that the theories make available for the description of phenomena; and (3) the values (accuracy, consistency, broad scope, **simplicity**, fruitfulness) that are used for the evaluation of **scientific theories**. *Exemplars* are model solutions to problems. They articulate the meaning of the fundamental concepts of the paradigm.

Further reading: Kuhn (1962)

Paradox: A sequence of claims such that, if they are taken in isolation from each other, they all seem reasonable and sound but taken together they lead to contradiction or to absurdity. Alternatively, an **argument** that appears to draw, by means of sound reasoning, a false conclusion from true premises. A paradox is resolved either by rejecting some of the premises or by challenging the reasoning that led to the conclusion.

Further reading: Sainsbury (1988)

Paradox of the ravens: A **paradox** of **confirmation**, which took its name from the example that **Hempel** used to illustrate it, namely, *all ravens are black*. There are three intuitively compelling principles of confirmation which cannot be jointly satisfied; hence the paradox. First, **Nicod**'s condition: a universal generalisation is confirmed by its positive instances. That all ravens are black is confirmed by the observation of black ravens. Second, the principle of equivalence: if a piece of evidence confirms a hypothesis, it also confirms its logically equivalent hypotheses. Third, the principle of relevant empirical investigation: hypotheses are confirmed by investigating empirically what they assert. Take the hypothesis (H): All ravens are black. The hypothesis (H′) *All non-black things are non-ravens* is logically equivalent to (H). A positive instance of *H′* is a white piece of chalk. Hence, by Nicod's condition, the observation of the white piece of chalk confirms *H′*. Hence, by the *principle of equivalence*, it also confirms *H*, that is, that all ravens are black. But then the principle of relevant empirical investigation is violated. The hypothesis that all ravens are black is confirmed not by examining the colour of ravens (or of any other birds) but by examining seemingly irrelevant objects (like pieces of chalk or red roses). So at least one of these three principles should be abandoned, if the paradox is to be avoided. Philosophers differ as to what principle they disfavour.

See **Confirmation, Bayesian theory of; Confirmation, Hempel's theory of**

Further reading: Hempel (1965)

Partial entailment: Relation between statements that report the observational evidence and statements expressing a hypothesis (or theory), articulated by **Carnap** in his system of **inductive logic**. The evidence (cast in observational statements) is supposed to confirm a hypothesis to the extent in which it partially entails this hypothesis, where partial entailment is analysed as a relation between the range of the **evidence** and the range of the hypothesis.

See **Probability, logical interpretation of**

Further reading: Carnap (1950b)

Particular: As opposed to **universals**, a particular is an individual. (An individual substance, according to **Aristotle**.) There can be concrete particulars, that is, individuals existing in **space** and **time** and satisfying the **laws of thinghood,** and abstract particulars, that is, individuals existing outside space and time. Concrete particulars are entities like chairs, tables and electrons, while abstract particulars are numbers and God (if he exists). Particulars are the logical subjects of which attributes can be predicated, but which cannot be predicated of anything else. Bare particulars are supposed to be the substrata on which **properties** inhere – whatever would be left over if all properties of an entity were taken away.

See **Nominalism; Properties**

Further reading: Armstrong (1989); Quinton (1973)

Pascal's wager: Argument for making rational decisions under uncertainty, put forward by Blaise Pascal (1632–1662), a famous French philosopher and mathematician. In *Pensées*, Pascal considered the issue of believing in God (and in particular in life after death). Given that the existence of God is uncertain, and given that the actions one may consider are believing that God exists and disbelieving that God exists, Pascal argued that believing that God exists is the most rational action one can perform. The structure of his argument depends on what we now call the maximisation of the expected utility of a decision, that is, the product of the **probability** that an event will happen times its expected value. Pascal takes it that, if God exists, the expected value of believing in God is infinite;

hence even if we thought that the probability that God exists is equal to the probability that God does not exist, the product of an infinite value with any finite number will be infinite. If, in contrast, God does not exist, the expected value of not believing in God is finite, hence the expected utility will be finite. As Pascal put it, 'if you win you win everything, if you lose you lose nothing'. This kind of **argument** has been contested on many grounds, not least because it presupposes that belief is a matter of the will. Be that as it may, Pascal's argument is perfectly generalisable. This type of argument was used by **Reichenbach** in his pragmatic vindication of **induction**.

See **James; Voluntarism**

Further reading: Hájek (2003)

Peirce, Charles Saunders (1839–1914): American philosopher, logician and scientist, founder of **pragmatism**. Two of his most influential works were: *The Fixation of Belief* (1877) and *How to Make our Ideas Clear* (1878). He took it that the meaning of an idea or **concept** lies in its practical consequences. His work on logic and reasoning led him to a tripartite division of modes of reasoning into deduction, **induction** and hypothesis. If we start with a **deductive argument** of the form D: {All As are B; a is A; therefore, a is B}, there are two ways to re-organise the premises and the conclusion: I: {a is A; a is B; therefore All As are B}; and H: {a is B; All As are B; therefore a is A}. Pattern I, which Pierce called **induction**, starts with some observations about a set of individuals and returns a generalisation over all individuals of a certain domain, while pattern H, which Peirce called hypothesis, starts with a particular known fact (a is B) and a generalisation (All As are B), and returns a conclusion about a particular fact (that a is A). Here is the example he used: given the premises 'All the beans from this bag are white' and 'These beans are white', one can draw the hypothetical conclusion that 'These beans are from this bag'. For Peirce, *H* and *I* correspond to two distinct modes of **ampliative inference**: induction classifies, whereas hypothesis explains. Later on, Peirce introduced the term **abduction** to cover hypothetical reasoning. Peirce took it that science aims at the truth, but thought that **truth** amounts to the final settlement of opinion. Science is an essentially collective enterprise, exercised by a community of inquirers who follow the **scientific method**. Peirce thought the true and the real are what will be agreed upon by the members of this community in the ideal limit of scientific inquiry.

See **James; Verificationism**
Further reading: Peirce (1957)

Pessimistic induction: Argument which aims to undercut the realist thesis that the best **explanation** of the successes of current **scientific theories** is that they are truthlike. It is based on the claim that the history of science is replete with theories that were once considered to be empirically successful and fruitful, but which turned out to be false and were abandoned. If, the **argument** goes on, the history of science is the wasteland of aborted 'best theoretical explanations' of the **evidence**, it might well be that current best explanatory theories will take the route to this wasteland in due course, despite their empirical successes. In its original formulation, due to **Laudan**, this pessimistic conclusion was supposed to be inductively established based on a list of past successful-yet-false scientific theories. But it is best seen as a *reductio ad absurdum* of the realist thesis that current successful theories are truthlike. The argument does not directly imply that current successful theories are false. Its aim is to discredit the claim that there is an *explanatory connection* between empirical success and **truthlikeness**. Among the realist attempts to rebut the pessimistic induction, one that has become prominent is based on the point that theory-change is not as radical and discontinuous as the pessimistic induction presupposed. Realists have aimed to show that there are ways to identify the theoretical constituents of abandoned scientific theories that essentially contributed to their successes, separate them from others that were 'idle', and demonstrate that it was those components which made essential contributions to the theory's empirical success that were retained in subsequent theories of the same domain.

See **Scientific realism; Structural realism**

Further reading: Kitcher (1993); Ladyman (2002); Psillos (1999)

Phenomenalism see **Sense data**

Physicalism: Philosophical doctrine committed to the following two theses: (1) the world as it is independently of us humans is a physical world; and (2) all facts are, ultimately, physical facts. Physicalism is the strongest version of **naturalism** in that it takes physical facts to be those described in the language of physics while naturalism is more open-minded: it allows that there are natural facts that are not reducible to physical facts. In a nutshell, physicalism is the view that the physical world is all there is. Physicalism faces two significant problems. One is *motivational*: why should one want to be physicalist in the first place? The other is *justificatory*: is physicalism a defensible position at all? Opponents of physicalism argue that it is not clear what physicalists mean when they talk about *the* physical. If the physicalist claim is that the realm of physical entities or facts is circumscribed by the semantic values of the physical vocabulary (i.e., the vocabulary of the physical sciences), given that the boundaries of this vocabulary are not clear-cut, nor eternally fixed, physicalism becomes an empty doctrine. They also claim that all attempts to reduce non-physical vocabulary to the physical (assuming, for the sake of the argument, that this is fixed) have failed. Finally, they claim that, even if every object that exists is a physical object, it does not follow that every fact is a physical fact, since there may well be non-physical **properties**. The strongest argument for physicalism is based on the following premises. (1) Causal completeness of physics: the physical realm is causally complete, in the sense that if anything is a physical effect, it must have a physical cause. (2) The causal sufficiency of the physical: physical causes are fully sufficient to produce (or fix the chances of) physical effects. Denying this leads to causal interactionism or causal **overdetermination**: at least on some occasions, some non-physical causes will be necessary for physical effects or some physical effects will have both physical *and* non-physical causes. (1) and (2) imply that the realm of the physical admits of a full causal explanation in physical terms. An opponent of physicalism can always argue that non-physical facts are causally inert. This would be a kind of epiphenomenalism about the non-physical. Hence, the argument for physicalism is that physicalism is the only doctrine that captures the causal give-and-takes in the world without giving rise to implausible philosophical theses such as epiphenomenalism, interactionism and causal overdetermination.

See **Reduction**

Further reading: Papineau (1993); Poland (1994)

Platonism, mathematical: The view that there are numbers, *qua* abstract entities, and that knowledge of them is possible. That there are numbers as determinate objects is required for the **truth** of arithmetical statements. That they are *abstract* objects is supposed to follow from the **literal interpretation** of arithmetical statements and, especially for **Frege**, from the failure of any other attempt to make sense of what numbers are. Typically, Platonists take the knowledge of numbers to be **a priori**. However, some (e.g., Frege) take the **knowledge** of numbers to be purely logical, while others (e.g., Goedel) claim that the knowledge of numbers is based on some kind of intuition or rational insight. Under the influence of **Quine**, some Platonists have assumed that the knowledge of numbers is broadly a posteriori and empirical since arithmetical truths are confirmed like any other truths by being part of our web of beliefs. Paul Benacerraf (born 1930) has issued two important challenges to Platonism. First: numbers cannot be said to be objects (with determinate identity conditions) since number-theory underdetermines what numbers are; there is *no* way we can fix the reference of number-words – numbers, for instance, can be identified with different set-theoretic constructions which lead to conflicting outcomes. Second: assuming that numbers are (abstract) objects, truths about them cannot be known, since, being outside **space** and **time**, numbers cannot enter into causal interactions with knowers – nor can there be other reliable methods for knowing them.

See **Fictionalism, mathematical**

Further reading: Colyvan (2001); Shapiro (1997)

Plausibility: A feature of a hypothesis on the basis of which the hypothesis is deemed intuitively acceptable before any empirical **evidence** for it is being sought. Many philosophers think that the very idea of scientific **inference** requires plausibility judgements. Since many competing hypotheses will fit the available data or will be consistent with background information, some of them must be excluded from consideration as being implausible; and the remaining hypotheses must be ranked in terms of degrees of plausibility. Criteria that have been used in plausibility rankings are **simplicity**, explanatory power, naturalness, fertility and lack of **ad hocness**. Though few doubt that scientists employ criteria of this sort in allotting initial plausibilities to competing hypotheses, many argue that they have no rational force; they express pragmatic considerations of conceptual economy. Others, however, claim that judgements of initial plausibility can be rational and objective since they are themselves evidential judgements: they encapsulate information about relevant background knowledge. Plausibility considerations have been taken to inform the assignment of prior probabilities to scientific hypotheses.

Further reading: Harman (1986); McMullin (1992)

Poincaré, Jules Henri (1854–1912): French philosopher and mathematician, author of *Science and Hypothesis* (1902), famous for his geometrical **conventionalism**. His work on the foundations of geometry, and in particular on the issue of the consistency of **non-Euclidean geometries**, led him to conclude that physical **space** is metrically amorphous and that one could take physical space to possess any geometry one liked, provided that one made suitable adjustments to one's physical theories. He suggested that the adoption of a certain geometrical system as *the* physical geometry was, by and large, a matter of **convention**. Poincaré argued that the axioms of **Euclidean geometry** are not empirical generalisations; nor are they **a priori** true, since one can imagine worlds in which non-Euclidean axioms hold. He called them 'conventions' (or, definitions-in-disguise). Poincaré extended his geometrical conventionalism further by arguing that the principles of mechanics were conventions. His starting point was that the principles of mechanics were not a priori truths, since they could not be known independently of experience. Nor were they generalisations of experimental facts. But calling the principles of geometry and mechanics 'conventions' did not imply, for Poincaré, that their adoption (or choice) was arbitrary. He stressed that some principles were more convenient than others. He thought that considerations of **simplicity** and unity could and should 'guide' the relevant choice. He envisaged a certain hierarchy of the sciences, according to which the very possibility of empirical and testable physical science requires that there are in place (as, in the end, freely chosen conventions) the axioms of Euclidean geometry and the principles of Newtonian mechanics. Yet, he thought that scientific hypotheses proper, even high-level ones such as Maxwell's laws, were empirical. Faced with the problem of discontinuity in theory-change, he argued that there is some substantial continuity at the level of the mathematical equations that represent empirical as well as theoretical relations. By and large, he thought, the theoretical content of scientific theories is structural: if successful, a theory represents correctly the *structure* of the world.

See **Structural realism**

Further reading: Poincaré (1902); Zahar (2001)

Popper, Karl Raimund (1902–1994): Austrian philosopher of science, who spent most of his academic career in the London School of Economics. He published a number of books, including *The Logic of Scientific Discovery* (1959) and *Objective Knowledge* (1972). He is mostly known for his critique of **inductivism** (Popper famously claimed that **induction** is a myth) and his defence of **falsificationism**. He developed a deductivist account of science and claimed that the basic methodological choices are conventional. His own method of **conjectures and refutations**

was supposed to highlight the difference between thinking of a hypothesis and subjecting it to severe testing. Though he took it that there is a sharp dictinction between science and non-science (based on the idea of falsifiability), he claimed that **scientific theories** emerge as attempts to articulate and render testable metaphysical programmes about the structure of the physical world. In his attempt to offer an objectivist account of **knowledge**, he drew a distinction between three worlds: the physical world (world 1); the (subjective) psychological world (world 2); and the world of ideas (world 3), that is, the world of the *logical content* of thoughts, books, computer memories and the like. This world 3 was taken to be the world were objective knowledge resides. This move was supposed to dissociate knowledge from the (subjectivist) state of **belief**. Popper was right when he stressed that knowledge does not require **certainty** but wrong when he tried to dissociate knowledge from **justification** – and in particular from having (inductive) reasons to believe that something is true.

See **Corroboration; Critical rationalism; Demarcation, problem of; Probability, propensity interpretation of; Scientific method; Verisimilitude**

Further reading: Miller (1994); Popper (1959, 1963)

Positivism: Originally, philosophical doctrine advanced by the French thinker Auguste Comte (1798–1857). It laid emphasis on reason and logic in searching for the facts and in coming to accept any theories. It stressed that theories should be licensed (Comte demanded that they should be *demonstrated*) by observations and arguments from **analogy**. It also stressed the indispensability of inductive reasoning especially when it came to first principles, since, as Comte put it, 'a principle which is the basis of all deduction cannot be itself deduced'. In the twentieth century, it was associated with the philosophy of the **Vienna Circle**, known as **logical positivism**. The British philosopher A. J. Ayer (1910–1989) presented and defended many of the central doctrines of the positivist movement in his *Language, Truth and Logic* (1936).

Further reading: Comte (1913); Giere and Richardson (1996)

Post hoc, ergo propter hoc: Fallacy in causal reasoning, meaning: after this, therefore because of this. It is a **fallacy** to claim that since *b* follows *a*, *b* is caused by *a*. Mere temporal succession is not a sufficient condition for causal connection. For instance, just because I tripped *after* a black cat crossed my path, it does not mean that I tripped *because* I encountered a black cat.

Further reading: Engel (2000)

Powers: The sources of activity in nature. According to a view that goes back to **Aristotle** and **Leibniz** and has recently resurfaced, **properties** are powers: active causal agents that are identified via their causal role (what kinds of effects they can produce). Two seemingly distinct properties that have exactly the same powers are, in fact, one and the same property. Similarly, one cannot ascribe different powers to a property without changing this very property. **Harré** and Edward H. Madden (born 1925), who were among the first to re-introduce powers into contemporary philosophical thinking, have drawn a distinction between Aristotelian individuals and Parmenidean ones. Aristotelian individuals have variable powers (i.e., powers that can change, fade away, die out etc.). This variability is grounded in the natures of these individuals: their nature can remain intact and yet their powers may change. Parmenidean individuals have constant powers and this constancy is constitutive of their nature: the powers and the nature of Parmenidean individuals are the same. Elementary particles, such as the electron and its constant power of negative charge, are taken to be examples of Parmenidean individuals. An increasingly popular claim is that at least *some* properties are *ungrounded* pure powers; that is, they are not grounded either in categorical properties or in other (more fundamental) powers. The fundamental properties of mass, spin and charge are taken to be ungrounded powers. Some philosophers – notably George Molnar (1934–1999) – think that the mark of powers is that they possess physical intentionality: they are directed towards their characteristic manifestation. Modern critics of powers offer an analysis of powers in terms of subjunctive and **counterfactual conditionals**: if *so-and-so* were the case, then power *F* would be exercised. They further suggest that the thought that all properties are powers, far from grounding the presence of activity in nature, fails to explain it. If properties are nothing but powers, then, when a power is manifested, its *effect* (the acquiring of a property by a particular) will also be a *power*. Hence, nothing really happens apart from the shifting around of powers from **particular** to particular. The presence of activity in nature is then accounted for by reference to the **laws of nature**.

See **Dispositions**

Further reading: Harré and Madden (1975); Shoemaker (1984)

Pragmatism: Philosophical school that has shaped most of philosophy in the USA. Its three most important defenders were **James, Peirce** and John Dewey (1859–1952). It has influenced the thought of philosophers as diverse as **Quine, Sellars** and Richard Rorty (born 1931). Though it is hard to offer a succinct characterisation of its basic tenets, it can be safely said that it focuses on practice as opposed to theory and claims that success in practice is the final arbiter of truth. *Qua* a theory of meaning, pragmatism stresses that the meaning of a statement (or a whole doctrine) consists in its practical consequences and in particular in the difference its truth makes in experience. *Qua* a theory of **truth**, pragmatism (as advocated by James) suggests that the truth is whatever works, while pragmatism (as advocated by Peirce) takes the truth to be the final settlement of opinion of inquirers (in the ideal limit of inquiry) after the constant application of the self-corrective *scientific method*. Dewey denounced the spectator theory of **knowledge** and argued, against uncritical **empiricism,** that our actions play a fundamental role in our understanding of the world. He also denounced **foundatonalism** and the quest for **certainty** in knowledge and claimed that whatever **beliefs** are warranted through human inquiry constitute knowledge.

Further reading: Rorty (1982)

Prediction vs accommodation: The accommodation of already known facts within a theory is sometimes contrasted to the prediction of hitherto unknown facts by it. Some philosophers think that prediction (understood as temporally **novel prediction**) counts more than accommodation (even if the latter is understood as a use-novel prediction). Others think that the provenance of the predicted phenomenon should make no difference to the support it lends to the theory that predicts it. For instance, one can imagine a case in which, unbeknown to the theoretician whose theory made the prediction of a novel phenomenon, the phenomenon had been already discovered by some experimenter. Would or should this information affect the support which the predicted fact confers on the theory? The relevant intuitions are not compelling, but many argue that insofar as the theoretician had not used information relevant to this phenomenon in the construction of the theory, or in so far as the theoretician had not 'cooked up' the theory to accommodate a phenomenon, whether or not the phenomenon was known should make no difference to the support it lends to the theory that predicts it. Accordingly, many argue that the real contrast is between prediction and ad hoc accommodation of a known fact. But it is also fair to note that prediction of hitherto unknown facts carries an additional weight vis-à-vis the **confirmation** of a theory because a theory that predicts a hitherto unknown phenomenon takes an extra risk of refutation.

Further reading: Maher (1993)

Preface paradox: The kind of **paradox** that arises when the author of a book, who is ready to assert everything that the books says, writes a disclaimer in the preface of her book saying that it is likely that there are errors or mistakes in the book. This situation is paradoxical because the author asserts each and every statement of the book (say $S_1, S_2, \ldots, S_n$) and at the same time (in the preface) she claims that not all of these statements are true (i.e., that it is not the case that S_1 and S_2 and ... and S_n). The paradox reveals a difference between two kinds of **evidence** we may have for our **beliefs**: first-order evidence for the **truth** of some beliefs we hold and second-order evidence for the claim that in the past we have been wrong about some of the beliefs we have had, despite the fact that we have had some first-order evidence for them.

Further reading: Sainsbury (1988)

Primary vs secondary qualities see **Berkeley; Galileo; Locke**

Principal principle: Methodological principle, defended by **Lewis,** according to which the subjective **degree of belief**

in an event A happening given that the **chance** of A is p should also be p. For instance, the subjective **probability** that a coin toss will land tails, given that the objective chance of landing tails is $1/2$, is also $1/2$. This principle is taken as a constraint that subjective (or epistemic) probabilities must satisfy as well as an implicit **definition** of objective probabilities.

See **Probability, propensity interpretation of; Probability, subjective interpretation of**

Further reading: Lewis (1980)

Principle of acquaintance: Enunciated by **Russell**, it states that every proposition that we can understand must be composed of ingredients with which we are acquainted. Acquaintance requires that the meaning of a word is given immediately in experience (e.g., by means of ostention). Consequently, the meanings of words which purport to refer to entities with which we cannot be acquainted should be defined in terms of words whose own meaning is directly given in experience.

See **Concept empiricism; Definition**

Further reading: Russell (1912)

Principle of indifference: A rule for assigning probabilities. Suppose that there are $n > 1$ exclusive and exhaustive possibilities (e.g., six possible outcomes of tossing a fair die). The principle says that each of them should be assigned a probability equal to $\frac{1}{n}$ ($\frac{1}{6}$ in the case of the die). In its epistemic version, the principle states that if there is no reason to believe that one possibility is more likely to turn up compared to the others, equal probabilities should be assigned to all of them. This principle, known also as the principle of insufficient reason, was the foundation of the logical interpretation of **probability**, as it was taken to be a logical principle. It played a central role in **Carnap's**

system of **inductive logic**. Though this principle has a lot of intuitive pull, it leads to a number of paradoxes. Depending on the parameters with which we describe a situation, it assigns different probabilities to outcomes. For instance, given a partition of discrete possibilities in terms of red and non-red, the probability of a book being red is a half. But, given a finer partition of possible colours (e.g., red, green, blue and pink) the probability that the book is red is one quarter. Keynes modified the classical principle of indifference attributing equal prior probabilities to the not-further-decomposable constituents of a series of alternatives. But even this modification leads to **paradox**.

See **Induction, the problem of**; **Probability, classical interpretation of**

Further reading: Carnap (1950b); Howson and Urbach (2006); Keynes (1921)

Principle of induction: It asserts the following: (1) the greater the number of cases in which *A* has been found associated with *B* in the past, the more *probable* it is that *A* is always associated with *B* (if no instance is known of *A* not associated with *B*); (2) a sufficient number of cases of association between *A* and *B* will make it nearly certain that *A* is always associated with *B*. Thus stated, the principle of induction *cannot* be refuted by experience, even if an *A* is actually found *not* to be followed by a *B*. But neither can it be proved on the basis of experience. **Russell**, who introduced this principle, took it to be a synthetic **a priori** principle. His claim was that without a principle like this science is impossible and that this principle should be accepted on the ground of its intrinsic **evidence**. But, as Keynes observed, Russell's principle of induction requires the **principle of limited variety**. Though synthetic, this last principle is hardly **a priori**.

See **Induction, the problem of**
Further reading: Russell (1912)

Principle of limited variety: Introduced by Keynes as a requirement for inductive **inference**. Suppose that although C has been invariably associated with E in the past, there is an unlimited variety of **properties** $E_1, \ldots, E_n$ such that it is logically possible that future occurrences of C will be accompanied by any of the E_is, instead of E. Then, and if we let n (the variety index) tend to infinity, we cannot even start to say how likely it is that E will occur given C, and the past association of Cs with Es. The principle of limited variety excludes the possibility just envisaged.

Further reading: Keynes (1921)

Principle of minimal mutilation: Methodological principle advanced by **Quine**. It favours conservatism in the revision of beliefs. When there is a need to revise our web of **belief**, Quine counsels us to make the minimal changes that are required for the restoration of coherence. This principle is compatible with the claim that even logical and mathematical truths can be abandoned in the light of recalcitrant **evidence** – the reason that they are not is that, since logic and mathematics are central to our web of belief, changing them would lead to a maximal mutilation of our web of belief.

Further reading: Quine and Ullian (1978)

Principle of tolerance: Advocated by **Carnap** in *The Logical Syntax of Language* (1934). '*It is not our business to set up prohibitions, but to arrive at conventions . . . In logic there are no morals.* Everyone is at liberty to build up his own logic, i.e. his own language, as he wishes. All that is required of him is that, if he wishes to discuss it, he must state his methods clearly, and give syntactical rules instead of philosophical arguments' (§17). Later, he called it the principle of conventionality of language forms. Given that the choice of language is a conventional matter (to be evaluated only in terms of its practical fruitfulness), Carnap thought that the aim of the philosophy of science was to make clear the different language *forms* that are adopted by rival parties in several philosophical and scientific disputes. Far from being genuinely factual, these disputes, Carnap thought, centre around suitable choices of languages. The principle of tolerance is thus part of Carnap's attempt to eliminate so-called metaphysical **pseudo-problems** from the sciences. It formulates a metatheoretical standpoint in which issues of ontology are replaced by issues concerning logical syntax.

See **Conventionalism**

Further reading: Carnap (1934)

Principle of uniformity of nature: In its classic formulation (due to **Hume**), it asserts that the course of nature continues always uniformly the same; that is, that the empirical regularities that have been discovered to hold so far will hold in the future. It has been argued that this principle is required for the **justification** of **induction**. Yet, this very principle is neither demonstratively true (changes in the course of nature can always be envisaged) nor empirically justifiable (since any attempt to justify it empirically would rest on an inductive argument). Hence, it has been claimed that any attempt to ground induction on this principle will be question-begging and circular. **Mill** took this principle to be based on a second-order induction over first-order regularities: the uniformity of nature was supposed to be resolved into the regularities that there have been found to be present in the phenomena. In spite of all this, the problem is that nature is *not* uniform in all respects.

See **Induction, the problem of**

Further reading: Mill (1911)

Probability: Mathematical theory first introduced in the seventeenth century in connection with games of chance and fully axiomatised by the Russian mathematician Andrei Nikolaevich Kolmogorov (1903–1987) in *Foundations of the Theory of Probability* (1933). Apart from its use in the sciences, probability theory has become very important to the philosophy of science especially in relation to the theories of **confirmation** and **induction**. Though there has been little disagreement in relation to the mathematical formalism, there has been considerable controversy regarding the interpretation of the formalism – and in particular the *meaning* of the concept of probability. There have been two broad strands in understanding probability: an epistemic and a physical. According to the first, probability is connected with **knowledge** or **belief** in that it expresses degrees of knowledge, or degrees of belief, or degrees of rational belief. According to the second strand, probabilities, like masses and charges, are objective features of the world. The epistemic strand is divided into two camps according to whether probabilities express a rational (objective) or merely a subjective **degree of belief**. Both camps agree that the probability calculus is a kind of extension of ordinary deductive logic, but the subjectivists deny that there are logical or quasi-logical principles (like the **principle of indifference**) which ought to govern the rational distribution of prior probabilities. The physical strand is divided into two camps according to whether or not there can be irreducible single-case probabilities (or chances). The advocates of the view that probabilities are relative frequencies take the concept of probability to be meaningful only if it is applied to a collective of events, while the advocates of **chance** take it to be meaningful that probabilities can be attributed to single unrepeated events. Historically, the epistemic conception of probability came first, as exemplified in the classical interpretation, while the conception of physical probability was developed as a reaction to the continental rationalism of **Laplace** and his followers. Richard von Mises (1883–1953), who was one of the founders of the view that probabilities are limiting relative frequencies, argued that probability theory was an empirical science (like mechanics and geometry) which deals with mass phenomena (e.g., the behaviour of the molecules of a gas) or repetitive events (e.g., coin tosses). He then tried to develop the theory of probability on the basis of empirical laws, namely, the law of stability of relative frequencies and the law of randomness. **Carnap** aimed to bring together the epistemic and the physical strands under his two-concept view of probability.

See **Bayesianism; Confirmation; Explication**

Further reading: Earman (1992); Gillies (2000); Howson and Urbach (2006); Skyrms (2000)

Probability, classical interpretation of: Advocated by most of the founders of the probability calculus and most notably by **Laplace**, it defines **probability** as the ratio of favourable to equally possible instances. For instance, the probability that a fair coin will land tails in a toss is the ratio of the number of favourable instances (tails) over the number of all equally possible instances (heads, tails); that is, it is one half. The classical interpretation takes it that the probabilities are measures of *ignorance*, since an equal possibility of occurrence is taken to imply that there is no reason to favour one possible outcome over the others. The principle that operates behind the classical interpretation is the **principle of indifference**.

Further reading: Carnap (1950b); Laplace (1814)

Probability, frequency interpretation of: According to this view, **probability** is about the relative frequency of an attribute in a collective. As von Mises (one of the founders of this interpretation) put it, first the collective, then the probability. A collective is a large group of repeated events, for example, a sequence of coin tosses. Suppose that in such a collective of n tosses, we have m heads. The *relative frequency* of heads is m/n. Probability is then defined as limiting relative frequency, that is, the limit of the relative frequency m/n, as n tends to infinity. A consequence of this account is either that probabilities cannot be meaningfully applied to single events or that to attribute some probability to a single event is to transfer a probability associated with an infinite sequence to a member of this sequence. Being **properties** of sequences of **events**, probabilities become fully objective. There is no guarantee, of course, that the limit of the relative frequency exists. That relative frequencies converge in the limit is a postulate of the frequency interpretation. And even if the limit of the relative frequency does exist, any finite relative frequency might well be arbitrary far from the limiting relative frequency. Hence, making short-term predictions on the basis of actual relative frequencies is hazardous. What is guaranteed is that *if* the limit of the relative frequency does exist, there will be convergence of the actual relative frequencies on it *in the long run*. This fact has been used by **Reichenbach** (another leading advocate of the frequency interpretation) in his attempt to offer a pragmatic vindication of **induction** – and in particular of the **straight rule of induction**. It turns out, however, that this property of convergence is possessed by any of a class of asymptotic rules, which nonetheless yield very different predictions in the short run.

See **Induction, the problem of; Laplace**

Further reading: Reichenbach (1949); von Mises (1957)

Probability, inductive: Typically, a concept applied to arguments. The inductive probability of an **argument** is the probability that its conclusion is true given that its premises are true. The inductive probability of an argument is a measure of the inductive strength of the argument, namely, of how strongly the premises support its conclusion. A strong inductive argument has high inductive probability.

See **Inductive logic**

Further reading: Carnap (1950b); Skyrms (2000)

Probability, logical interpretation of: It conceives of **probability** as a logical relation that holds between propositions. This logical relation is taken to be a relation of **partial entailment** – for instance, it is said that though the proposition p does not (deductively) entail the conjunction (p and q), it entails it partially, since it entails one of its conjuncts (namely, p). The probability calculus is then used to calculate the probability of a proposition (say, a hypothesis) in relation to another proposition (say, a proposition expressing the **evidence**) that partially entails it. On this approach, the degree of partial entailment is the degree of *rational* belief: the **degree of belief** a rational agent ought to have in the **truth** of the hypothesis in the light of the evidence that confirms it. This interpretation was defended by Keynes and **Carnap**. Keynes claimed that rational agents possess a kind of logical intuition by means of which they *see* the logical relation between the evidence and the hypothesis. But **Ramsey** objected to Keynes's claim that he could not see these logical relations and that he expected to be persuaded by argument that they exist. Carnap developed the logical interpretation into a quantitative system of **inductive logic**.

Further reading: Carnap (1950b); Keynes (1921)

Probability, posterior: The **probability** of a hypothesis *given* some **evidence** (or, after evidence or other relevant information has been collected); hence, the **conditional probability** of the hypothesis given the evidence. The difference, if any, between the posterior probability and the prior probability of a hypothesis captures the degree of **confirmation** of the hypothesis. The relation between posterior probability and prior probability is given by **Bayes's Theorem**.

Further reading: Howson and Urbach (2006)

Probability, prior: The **probability** of a hypothesis (or the probability that an event will occur) before any **evidence** (or relevant information) is collected. For instance, the prior probability that a fair die will land six on a toss is one-sixth. A major issue in the philosophy of probability and in **confirmation** theory is exactly how prior probabilities are determined and what status they have (whether, in particular, they are subjective or objective-logical **degrees of belief**).

See **Bayesianism**

Further reading: Howson and Urbach (2006)

Probability, propensity interpretation of: It takes probabilities to be objective **properties** of single (and unrepeated) **events**. In the version defended by **Popper**, propensities are properties of experimental conditions (**chance** set-ups). Hence, a fair coin does *not* have an inherent propensity of one-half to land heads. If the tossing takes place in a set-up in which there are slots on the floor, the propensity of landing heads is one-third, since there is the (third) possibility that the coin sticks in the slot. This account of **probability** is supposed to avoid a number of problems faced by the frequency interpretation. In particular, it avoids the problem of inferring probabilities in the limit. But, especially in Popper's version, it faces the problem of specifying the conditions on the basis of which propensities are calculated. Given that an event can be part of widely different conditions, its propensity will vary according to the conditions. Does it then make sense to talk about *the* true objective singular probability of an event? In any case, it has been argued that probabilities cannot be identified with propensities. The problem is this. There are the so-called inverse probabilities, but it does not make sense to talk about inverse propensities. Suppose, for instance, that a factory produces red socks and blue socks and uses two machines (Red and Blue) one for each colour. Suppose also that some socks are faulty and that each machine has a definite probability to produce a faulty sock, say one out of ten socks produced by the Red machine are faulty. We can meaningfully say that the Red machine has an one-tenth **propensity** to produce faulty socks. And we can also ask the question: given an arbitrary faulty sock, what is the probability that it has been produced by the Red machine? This question is meaningful and has a definite answer. But we cannot make sense of this answer under the propensity interpretation. We cannot meaningfully ask: what is the propensity of an arbitrary faulty sock to have been produced by the Red machine?

See **Probability, frequency interpretation of**

Further reading: Humphreys (1989); Popper (1959)

Probability, subjective interpretation of: Also known as subjective **Bayesianism**, it takes probabilities as subjective degrees of belief. Opposing the objective or logical interpretation of **probability**, it denies that there is such thing as *the* rational degree of belief in the truth of a proposition. Each individual is taken to (or allowed to) have her own subjective **degree of belief** in the **truth** of a certain proposition. Given that the probability calculus does not establish any (prior) probability values, subjectivists argue that it is up to the agent to supply the probabilities. Then, the probability calculus, and **Bayes's theorem** in particular, is used to compute values of other probabilities based on the prior probability distribution that the agent has chosen. The only requirement imposed on a set of degrees of beliefs is that they are probabilistically coherent, that is, that they satisfy the axioms of the calculus. The rationale for this claim is the so-called **Dutch-book theorem**. Critics of subjectivism argue that more constraints should be placed on the choice of prior probabilities, but subjectivists counter that all that matters for **rationality** is how degree of beliefs hang together at a time and how they are updated over time, in the light of new **evidence**. Subjectivists also appeal to the **convergence of opinion** theorem to claim that, in the long run, prior probabilities wash out. But this is little consolation because, as Keynes put it, in the long run we are all dead.

See **Coherence, probabilistic; Probability, prior**

Further reading: Howson and Urbach (2006); Skyrms (2000)

Progress see **Conjectures and refutations; Kuhn; Lakatos**

Projectability see **Grue**

Propensity: Probabilistic **disposition** or tendency to behave in a certain way. It is taken to be an objective property of either an object (e.g., the propensity of a radioactive atom to decay) or a whole experimental condition (e.g., the propensity of a fair coin to land heads given that it is tossed on a surface with slots). Many philosophers take propensities to be irreducible features of the physical world – posited for theoretical/explanatory reasons. Others take the view that propensities are reducible to relative frequencies. Propensities are taken to be necessary for understanding physical **chance**, especially in **quantum mechanics**.

See **Probability, frequency interpretation of; Probability, propensity interpretation of**

Further reading: Humphreys (1989); Mellor (1991); Popper (1959)

Properties: Ways things are. To describe an object as being red or spherical or charged and the like is to ascribe a property to it. Traditionally, an object having (or possessing) a property is a minimal requirement for the existence of a states of affairs (or facts). Properties are taken to be the intensions of predicates, while classes of things (those that have the property) are the extensions of predicates. There are a number of philosophical disputes concerning properties.

1. Nominalists either deny the existence of properties altogether or try to reduce them to classes of particulars. Yet some take properties to be **particulars**, known as **tropes**. Realists admit the existence and ineliminability of properties. They take properties to be **universals**, but there is a division among those realists who claim that properties can exist only in things, and those realists who think that properties exist prior to and independently of things.

2. The modal status of properties: are they categorical or dispositional? Dispositionalists take properties to be (active and passive) **powers**, while categoricalists take properties to be purely qualitative and inert, trying to account for the presence of activity in nature on the basis of **laws of nature**.

3. Essential properties vs accidental ones: some philosophers take all properties to be on a par, while others take some properties to characterise their bearers either essentially or accidentally.

4. Properties vs particulars: do properties require some particular to bear them or are objects nothing more than bundles of properties? The advocates of the bundle view take properties to be the fundamental building blocks of nature and argue that particulars are constituted by compresent properties. Those who take both particulars *and* properties to be the fundamental building blocks of nature typically argue that a substratum on which properties inhere is necessary for understanding continuity and change in the life of an object.

5. Finally, are there natural properties? Though on an extensional account of properties any odd class of things can be a property (since it is a class), many philosophers argue that some properties are more natural than others since their bearers display some kind of objective similarity.

See **Dispositions; Essentialism; Natural kinds; Nominalism**

Further reading: Heil (2003)

Protocol sentences: Sentences that were supposed to act as the foundation of all scientific **knowledge**. They were introduced by the logical positivists and the issue of their status and content embroiled them in a heated debate in the beginning of the 1930s, known as the protocol sentences debate. The expression 'protocol sentences' was meant to capture the fact that they were registered in scientific *protocols*, which report the content of scientists' observations. Protocol sentences were understood in two different ways. They were taken either as expressible in a **sense data**-language, or as expressible in a thing-language. For instance, a protocol statement can have either the form 'here now blue', or 'A red cube is on the table'. But protocol sentences were not generally understood as observational statements proper, that is, as *standing* statements which express intersubjective results of observations. **Schlick**, for instance, conceived them as *occasion* sentences, that is, as tokens of observational statements expressible in the first person, upon the occasion of an observation or perception by a subject. **Carnap** toyed with the idea that protocol statements need no **justification**, for they constitute the simplest states of which knowledge can be had. But he was soon convinced by **Neurath** that there are neither primitive protocol sentences nor sentences which are not subject to verification. Instead of abandoning the claim that science provides knowledge on the grounds that this knowledge cannot be certain, Carnap opted for the view that scientific knowledge is short of **certainty**.

See **Foundationalism**

Further reading: Ayer (1959); Uebel (1992)

Pseudo-problems: Philosophical problems that appear to be genuine and to call for a deep philosophical solution but turn out to be senseless. **Carnap** and the logical positivists took most traditional philosophical problems (e.g., the problem of the existence of an external world) to be pseudo-problems: when properly analysed, they reduce to problems concerning the choice of linguistic frameworks. The judgement that they are pseudo-problems was based on the **verifiability** criterion of meaningfulness.

See **External/internal questions; Formal mode vs material mode; Principle of tolerance**

Further reading: Carnap (1928)

Pseudo-science see **Demarcation, problem of**

Putnam, Hilary (born 1926): One of the most influential American philosophers of the second half of the twentieth century, with ground-breaking contributions to many

areas of the philosophy of science, including mathematical logic and the foundations of artificial intelligence. He
is the author of *Meaning and the Moral Sciences* (1978)
and *Reason, Truth and History* (1981). He was a critic
of **logical positivism** and of **instrumentalism** and one of
the early defenders of the realist turn in the philosophy
of science. He challenged the **verifiability** theory of meaning and argued that theoretical **terms** get their meaning
holistically, by being part of theories. He argued that theoretical terms refer to **unobservable entities**, which are no
less real than observable ones. He advanced the explanationist defence of **scientific realism**, according to which
scientific realism is an overarching empirical hypothesis
that constitutes the best explanation of the success of science. He defended the reality of numbers – *qua* **abstract
entities** – based on what came to be known as the Quine–
Putnam indispensability argument. Though influenced by
Quine, he defended some version of the **analytic/synthetic
distinction**, arguing that some concepts are one-criterion
concepts, while others (including most scientific concepts)
are introduced via theory and by what he called 'property-
clusters'. In the 1970s, he defended and elaborated the
causal theory of reference and used it to defend semantic
externalism, the view that the meaning of a concept is not
fixed by the internal mental states of the speaker. Though
he was one of the inventors of functionalism in the philosophy of mind, the view that mental states are individuated
via their causal-functional role, he came to abandon this
view later on in his career. He also abandoned his robust
metaphysical realism, under the influence of **Dummett**. As
an alternative to metaphysical realism, Putnam developed
internal realism. This is the view that there is no 'God's
Eye point of view', that is, there is no one and unique true
description of the world. One of Putnam's metaphors is
the dough and the cookie-cutter: if the world is a piece

of dough, what kind of objects there are depend on the cookie-cutter one uses to carve up the dough – that is, on the conceptual scheme and the categories one employs. Recently, he has advocated **pragmatism** and some version of direct **realism**.

Further reading: Putnam (1978, 1981)

Q

Quantum mechanics, interpretations of: *Qua* a mathematical theory, quantum mechanics has been interpreted in many different ways. Each interpretation can be taken to yield a different theory; these theories are empirically equivalent, but explain the world according to different principles and mechanisms. The so-called *orthodox* interpretation, also known as the Copenhagen interpretation, goes back to the works of **Bohr** and Heisenberg. The basic claim behind this interpretation is that the wavefunction is subject to two distinct types of process: a deterministic evolution according to Schroedinger's equation; and a stochastic collapse of the wavefunction during the measurement. The Schroedinger evolution determines the probabilities that a certain quantum system (e.g., an electron) will be found to be in a certain state. According to the Copenhagen interpretation, there is no fact of the matter as to what state a quantum system is in in-between measurements; the collapse somehow *puts* the quantum system in a definite state – the one measured during the measurement. Alternative interpretations have mostly arisen out of dissatisfaction with the Copenhagen interpretation. A view, defended by David Bohm (1917–1992) (and Louis de Broglie, 1892–1987, with his ingenious idea of a pilot-wave) was that quantum mechanics is incomplete: there are others parameters (known as *hidden* *variables*) such that, when they are taken into account, they determine the state of the quantum system. Hence, the description of the quantum system offered by the wavefunction is completed by the specification of extra parameters (e.g., the positions of the particles). This idea, that quantum mechanics is incomplete, was suggested in a famous paper, entitled 'Can Quantum-Mechanical Description of Physical Reality Be Considered Complete?' (1935), by **Einstein**, Boris Podolsky and Nathan Rosen. They argued for the claim that the wavefunction does not provide a complete description of the physical reality and devised a famous **thought experiment** to prove it. The gist of this was that one could determine (predict) with certainty the value of a parameter A of a system S by making a measurement of a value B of a correlated system S', where, though S and S' have interacted in the past, they are now so far apart from each other that they can no longer interact in any way. EPR (as the three authors of the foregoing paper came to be known) concluded that system S must already possess the predicted value and since the quantum mechanical description of the state of S fails to determine this, quantum mechanics must be incomplete. An assumption employed by EPR was the principle of locality (suggested by Einstein's theory of relativity), that is, the principle of no action-at-a-distance. John Stewart Bell (1928–1990), in his celebrated paper 'On the Einstein–Podolsky–Rosen Paradox' (1964), proved that quantum mechanics violates locality. This suggested that *any* interpretation of quantum mechanics should be non-local. Bohmian mechanics is a non-local theory since it relies on a non-local quantum potential. In 'Relative State Formulation of Quantum Mechanics' (1957), Hugh Everett (1930–1982) denied that there was collapse of the wavefunction. According to his many worlds interpretation, it only *appears* that the superpositions collapse. This appearance is explained away by noting that every time in which an experiment on a quantum system is performed, *all* possible outcomes with non-zero probability obtain – but each of them in a different (parallel) world. And this is so irrespective of the fact that we are aware only of the outcome in the world we live. This feature has been described as world-splitting. Among the interpretations of quantum mechanics that admit the collapse of the wavefunction, two stand out. One has been put forward by Eugene Paul Wigner (1902–1995), who argued that the consciousness of the observer collapses the wavefunction. The other has been put forward by Gian Carlo Ghirardi, Alberto Rimini and T. Weber who argued that the collapse occurs via a new kind of physical interaction.

Further reading: Albert (1992); Lange (2002); Redhead (1987); Torretti (1999)

Quasi-realism: Version of **realism** (or anti-realism) developed by the British philosopher Simon Blackburn (born 1944). The main thought is that quasi-realists can 'earn the right' to talk about the **truth**-or-falsity of theories, without the concomitant commitments to a realist (mind-independent) ontology: the posited entities inhabit a 'projected' world.

Further reading: Blackburn (1993)

Quine, W. v. O. (1908–2000): American philosopher, perhaps the most influential American philosopher of the twentieth century. His books include *Word and Object* (1960) and *Pursuit of Truth* (1992). He blended **empiricism** with elements of **pragmatism**. He defended **naturalism**, which he took it to be characteristic of empiricism, and denied the possibility of **a priori** knowledge. In 'Truth by Convention' (1936), he repudiated the view that logic was a matter of **convention**. In 'Two Dogmas of

Empiricism' (1951), he went on to argue that the notion of analyticity is deeply problematic, since it requires a notion of cognitive synonymy (sameness of meaning) and there is no independent criterion of cognitive synonymy. Quine went as far as to question the very idea of the existence of meanings. In his work during the 1950s and 1960s, he advanced a holistic image of science, where there are no truths of special status (necessary or unrevisable). What matters, for Quine, is that a theory acquires its empirical content as a whole, by predicting observations and by being confronted with experience. Then, should the theory come to conflict with experience, *any* part of the theory may be abandoned, provided that the **principle of minimal mutilation** is satisfied. He put forward five virtues that a scientific theory should have: conservatism, generality, **simplicity**, refutability and modesty. But the methodological status of these virtues was left unclear. Naturalism, for Quine, licenses his **scientific realism**. He never doubted the existence of **unobservable entities** and took it that their positing is on a par with the positing of most ordinary physical objects. They are both indispensable in formulating laws, the ultimate **evidence** for which is based on past experience and the prediction of future events.

See **Analytic/synthetic distinction; Convention; Duhem–Quine thesis; Neurath's boat; Platonism, mathematical; Underdetermination of theories by evidence; Universals**

Further reading: Orenstein (2002); Quine (1960)

Ramsey, Frank Plumpton (1903–1930): Perhaps the greatest British philosopher of the twentieth century. He published very little during his short life, but both his published work as well as his posthumously published papers and notes have exerted enormous influence on a number of philosophers and have set the agenda for many philosophical problems. In philosophy of science, he is mostly well known for his work on the structure of theories (especially through the so-called **Ramsey-sentences**); on **laws of nature** (through his defence of what came to be known as the Mill–Ramsey–Lewis approach to laws) and on the philosophy of **probability** (especially through his critique of Keynes's logical interpretation of probability and his advancement of the subjective interpretation). He advocated a deflationary approach to **truth** (known as redundancy theory of truth); he denied the distinction between **particulars** and **universals**, arguing that it is an artefact of language. He also defended the view that **knowledge** is reliably produced true **belief**.

See **Bayesianism; Probability, logical interpretation of**
Further reading: Ramsey (1931); Sahlin (1990)

Ramsey-sentences: To get the Ramsey-sentence of a (finitely axiomatisable) theory we conjoin the axioms of the theory in a single sentence, replace all theoretical predicates with distinct variables, and then bind these variables by placing an equal number of existential quantifiers in front of the resulting formula. Suppose that the theory TC is represented as TC $(t_1 \ldots t_n; o_1 \ldots o_m)$, where TC is a purely logical $m + n$-predicate. The Ramsey-sentence ^{R}TC of TC is: $\exists u_1 \exists u_2 \ldots \exists u_n$ TC $(u_1 \ldots u_n; o_1 \ldots o_m)$. Ramsey-sentences have a number of important properties. For instance, the Ramsey-sentence of a theory has exactly the same first-order observational consequences as the theory. Or, if two Ramsey-sentences are compatible with the same observational truths, they are compatible with each other. Ramsey-sentences are called thus because they were first introduced by **Ramsey** in his posthumously published essay 'Theories'. His critical insight was that the excess (theoretical) content of a theory over and above its observational content is seen when the theory is formulated as expressing an existential judgement of the foregoing form: *there are* entities which satisfy the theory.

See **Analytic/synthetic distinction; Carnap; Lewis; Maxwell, Grover; Structural realism**

Further reading: Carnap (1974); Psillos (1999); Ramsey (1931)

Rationalism: The view that reason alone (unaided by experience) can come to know substantive truths about the world. Hence, the view that there can be a priori knowledge of the world – of its basic laws, or structure. In the history of philosophy, it is associated with **Descartes**, Benedict de Spinoza (1632–1677) and **Leibniz**. Deduction from first principles is a key way to gain knowledge, according to rationalists. But the first principles themselves are known either intuitively or by rational insight. Rationalists do not deny the possibility of empirical science. They have aimed to ground science on indubitable and necessary truths that provide the foundation of all **knowledge**. Historically, rationalism has been associated with the view that there are innate ideas.

See **A priori/a posteriori; Certainty; Concept empiricism; Empiricism; Kant**

Further reading: Cottingham 1984

Rationality: Normative concept that characterises beliefs or actions. It has to do with reasons and reliability. A **belief** is rational if it is supported by reasons and, in particular, reasons that render this belief likely to be true. Equally a belief is rational if it is produced by reliable methods, that is, methods that tend to produce true beliefs.

How exactly reasons and reliability are connected is a tough issue. The appeal to reasons implies that attributing rationality to a belief amounts to attributing some cognitive virtue to the subject of the belief: the subject is rational because she/he is attentive to reasons. But the appeal to reliability implies that attributing rationality to a belief amounts to attributing some objective property to a method or cognitive process: the subject of the belief need not have reasons to think that the methods or cognitive processes she/he follows are reliable. As Robert Nozick (1938–2002) put it, reasons without reliability seem empty and reliability without reasons seems blind. The rationality of action is taken to be a means-end issue: rational action consists in following the best strategy that will promote one's aims. Rationality thus becomes goal directed, but the goals themselves are typically taken to be beyond rational adjudication. This purely instrumental conception of rationality can also be attributed to the rationality of belief. It may be said that the goal to which a (rational) belief is directed is **truth** (or some other cognitive virtue). This may be right to some extent, but there is a sense in which the rationality of a belief is *not* an instrumental property of this belief. Rather it is a function of the epistemic relation between the **evidence** and the belief for which it is taken to be evidence, and hence a function of the soundness of the methods that produced and sustain these beliefs. The rationality of science is typically associated with the **scientific method** and its **justification**. The defenders of the rationality of science are divided into two broad camps: those that take it that the scientific method needs justification and can be justified as a means for substantive **knowledge** of the world; and those who take it that the scientific method is a logical form empty of content (be it, deductive logic, **inductive logic**, or Bayesian **conditionalisation**), thereby restricting the rationality of science to how beliefs get connected to each other at a time and change over time.

See **Bayesianism; Critical rationalism; Feigl; Laudan; Reliabilism**

Further reading: Nozick (1993)

Realism and anti-realism: Historically, realism was a doctrine about the independent and complete existence of **universals (properties)**. It was opposed to **nominalism.** Currently, realism has a more general meaning. It affirms the objective **reality** (existence) of a class of entities, and stresses that these entities are mind-independent. Realism is primarily a metaphysical thesis. But many philosophers think that it has a semantic as well as an epistemic component. The semantic thesis claims that a certain discourse or class of propositions (e.g., about theoretical entities, or numbers, or morals) should be taken at face value (literally), as purporting to refer to real entities. The epistemic thesis suggests that there are reasons to believe that the entities posited exist and that the propositions about them are true. Given this epistemic thesis, realism is opposed to **scepticism** about a contested class of entities. *Anti-realism* can take several forms. One of them is anti-factualism: it understands the contested propositions (e.g., about unobservable entities, or mental states, or numbers) literally but denies that there are facts that make them true. Hence, it takes the contested propositions to be false and denies that there are entities that these propositions refer to. Mathematical **fictionalism** and ethical **error-theory** are species of this form of anti-realism. Another form of anti-realism is non-factualism: the 'propositions' of the contested class are not *really* propositions; they are not apt for truth and falsity; they are not in the business of describing facts. **Instrumentalism**, ethical noncognitivism and mathematical formalism are cases of this kind of anti-realism.

A third popular form of anti-realism comes from **Dummett** who argues that the concept of **truth** is epistemically constrained. Dummettian anti-realism does not deny that the contested propositions (e.g., about numbers) can be (and in fact are) true, but argues that their truth cannot outrun the possibility of verification. This species of anti-realism equates truth with warranted assertibility. Mathematical intuitionism is a case of this form of anti-realism. Given this Dummettian perspective, realism is the view that every proposition of the contested class is either true or false, independently of anyone's ability to verify or recognise its truth or falsity. Hence, realism is taken to subscribe to the logical principle of bivalence.

See **Scientific realism**

Further reading: Devitt (1997); Wright (1992)

Reality: Everything there is. The philosophical issue over reality concerns (1) its *scope*: what the elements (or constituents) of reality are and (2) its *status*: whether it exists independently of the human mind. For instance, philosophers have debated whether there are **universals** as opposed to **particulars,** whether there are material objects as opposed to **sense data,** whether there are **abstract entities** etc. They have also debated whether things can exist while unperceived, whether the world would be there even if there were no minds (or God) to think of it etc. Reality is also contrasted to appearances – to how things appear to perceivers. It is said to be independent of the appearances in that the reality could be there even if there were no appearances and in that how things *really* are can be fundamentally different from how they appear.

See **Idealism; Realism and anti-realism**

Redhead, Michael (born 1929): British philosopher of physics, author of *Incompleteness, Nonlocality and*

Realism (1987). He has worked on the compatibility between **quantum mechanics** and relativity theory and has criticised a simple-minded realism of possessed values. More recently, he has tried to defend some version of **structural realism**.

Further reading: Redhead (1987)

Reduction: According to an influential model advanced by **Nagel**, the reduction of a theory T to another theory T' requires two things. First, the vocabulary of T is suitably connected to the vocabulary of T'. This is what Nagel called 'connectibility axioms' (also known as bridge laws). So, if S is a concept of theory T and P is a concept of theory T', the bridge law should be a bi-conditional of the form S if and only if P. Second, T-sentences (that is, the sentences of theory T) are derivable from T'-sentences plus bridge laws. The idea behind this requirement is that reduction requires that the law-statements of the reduced theory T are proved to express laws of the reducing theory T'. The logical positivists, though advocates of the unity of language, were reticent of the idea that there is unity of laws. Nagel remained silent on the metaphysics of reduction. His model is fully consistent with the view that bridge laws stated either analytical **definitions** or mere correlations of predicates. For instance, a bridge law of the form 'everything that has colour has shape' can correlate colours and shapes but does not reduce colours to shapes. An alternative view was that the bridge laws were theoretical identifications, such as the identification of the temperature of a gas with the mean kinetic energy of its molecules. These identifications were taken to be a posteriori discoverable facts. Instead of taking the bridge laws to be brute facts, the theoretical identifications explained why they hold. In 'On the Unity of Science as a Working Hypothesis' (1958),

Putnam and Oppenheim favoured a micro-reduction of all objects to physical objects (ultimately, the elementary particles of physics) based on the part–whole relation: objects in the domains of the special sciences (biology, psychology etc.) are composed of objects in the domain of physics. Given a hierarchical organisation of all objects in levels (elementary particles, atoms, molecules, cells, organisms etc.), they argued that things at the higher level n + 1 are composed of things belonging to the lower level n and that, more strongly, things at each level have properties which are realised by **properties** of lower levels. Physical objects and properties are the ground zero of all things and properties. The unity of laws is achieved by means of the fact that all higher-level properties (that feature in higher-level laws) are realised by physical properties. But the advances in the special sciences, their explanatory and predictive strengths and their empirical successes made it all the more difficult to argue *against* their autonomy from physics. In 'Special Sciences (or: The Disunity of Science as a Working Hypothesis)' (1974), **Fodor** argued that the ontic priority and generality of physics does not imply reductionism. The latter requires identity of properties, or type-identities. That is, it requires that each property (type) S of a special science is identical to a physical property (type) P. Fodor argued that a weaker form of **physicalism** (token-physicalism) is strong enough to secure the ontic priority of the physical (since each token of a special property is identical to a token of a physical property) but, at the same time, weak enough to allow for the autonomy of the special sciences (since the types that special sciences deal in are not identical to physical types). Even if we allow for bridge laws, we don't thereby get type-identity, since the bridge laws guarantee only that the relevant predicates are co-extensional and not that they pick out one and the same property. Token-physicalism gained extra purchase by the advent of functionalism in the philosophy of mind, which capitalised on the fact of multiple realisability, namely, the fact that higher-level properties are realised by different physical properties. But the chief argument against type-physicalism (that is, reductionism) was that the special sciences formulate proper laws; their laws connect **natural kinds**; these laws and kinds play an ineliminable explanatory and predictive role.

See *Ceteris paribus* **laws; Emergence; Supervenience; Unity of science**

Further reading: Batterman (2001); Fodor (1974); Nagel (1960); Putnam and Oppenheim (1958)

Reduction sentences: Introduced by **Carnap** in an attempt to show how the meaning of theoretical **concepts** can be partly specified (implicitly defined) by virtue of the meanings of observational concepts. The introduction of reduction sentences was a turning point in the empiricist approach to the meaning of theoretical concepts since it flagged the abandonment of the hope that theoretical concepts could be eliminated on semantic grounds. According to Carnap, the introduction of a theoretical term or predicate Q is achieved by the following *reductive pair*: *For all x (if S_1x then (if R_1x then Qx)) and For all x (if S_2x then (if R_2x then not-Qx))* (RP), in which S_1, S_2 describe experimental (test-) conditions and R_1, R_2 describe characteristic responses (possible experimental results). In the special case in which $S_1 = S_2 (= S)$ and $R_1 = not$-$R_2 (= R)$, the reduction pair (RP) assumes the form of the bilateral reductive sentence: *For all x (if Sx then (Qx if and only if Rx))* (RS). Suppose, for instance, that we want to introduce the concept TEMPERATURE OF C DEGREES CENTIGRADE by means of a reduction sentence. This will be: if the test-conditions S obtain (i.e., if we put object a in contact with a thermometer), then a has temperature of c degrees centigrade if and only if the characteristic response R obtains (i.e., if and only if the thermometer shows c degrees centigrade). The introduction of theoretical concepts by means of reduction sentences does *not* face the problems of explicit **definitions**. However, the reduction sentences do not eliminate the concept Q. For although they provide a necessary and a sufficient **condition** for Q, these two conditions do not coincide. Thus, the meaning of Q is not completely specified by virtue of observational concepts.

See **Concept empiricism; Definition, implicit**
Further reading: Carnap (1936)

Reductive empiricism: Form of **empiricism** adopted by some logical positivists during the early 1930s. It treats theoretical statements as being reducible to observational statements. Hence, it treats theoretical discourse (i.e., discourse that involves theoretical terms) as being disguised talk about observable entities and their actual (and possible) behaviour. Reductive empiricism is consistent with the claim that theoretical statements have truth-values, but it understands their truth-conditions reductively: they are fully translatable into an observational vocabulary. Though, then, it allows that theories might be true, it is not committed to the existence of **unobservable entities**. This ontological and semantic reduction was supposed to be effected by explicit **definitions**.

See **Literal interpretation; Scientific realism**
Further reading: Carnap (1936); Psillos (1999)

Reichenbach, Hans (1891–1953): German philosopher of science, member of the Berlin-based Society for Empirical Philosophy, which was closely associated with the **Vienna Circle**. In 1933 he fled Germany to the University of Istanbul and then, in 1939, he emigrated to the USA. He did profound work in the philosophy of physics and the general philosophy of science. His books include: *Relativity Theory and the A Priori* (1921), *The Philosophy of Space and Time* (1928) and *Experience and Prediction* (1938). In his early work, he distinguished between two elements in the Kantian claim that some statements are knowable a priori: that they are necessarily true and that they are constitutive of the object of **knowledge**. He rejected the first element arguing that principles that were taken to be necessarily true (e.g., the axioms of **Euclidean geometry**) were challenged and revised. But he thought the second element in the Kantian conception of the a priori was inescapable. He claimed that knowledge requires some principles of co-ordination, which connect some basic concepts with elements of reality. He was led to a relativised approach to the a priori: the principles of co-ordination are revisable but for each conceptual framework some such principles should be in place. These are a priori relative to the conceptual framework. Once this framework is in place, a theory is presented as an axiomatic system, whose basic axioms – the 'axioms of connection' – are empirical. Maxwell's laws, for instance, were taken to be such axioms of connection. Under the influence of **Schlick**, Reichenbach was led to adopt **conventionalism**, arguing that the choice of the geometry of physical **space** was conventional. He was one of the chief advocates of the liberalisation of the logical positivist criterion of cognitive significance, favouring the idea that confirmability (as opposed to strict **verifiability**) was enough for meaningfulness. He advanced the relative frequency interpretation of **probability** and argued that the **straight rule of induction** could be vindicated pragmatically. He defended the compatibility of **empiricism** and **scientific realism**. His idea was that even if we grant, as we should, that all factual knowledge starts with experience, its boundaries depend on the credentials of the methods employed. It is perfectly compatible with empiricism to adopt ampliative methods and to accept the existence of **unobservable entities** on their basis.

See **A priori/a posteriori; Causal process; Causation, direction of; Context of discovery vs context of justification; Induction, the problem of; Validation vs vindication**
Further reading: Reichenbach (1921, 1938)

Relativism: Cluster of views that deny absolute perspectives and universal points of view. It may be seen as the claim that normative judgements have no force outside a certain context or background or community or framework in that there is no meta-perspective within which all different contexts, backgrounds etc. can be placed and evaluated. Sometimes, this claim takes the form that there is no 'God's-eye view': no way in which one can rise above one's conceptual scheme and make judgements about it and about its relation to others instead of merely from within it. Accordingly, the best one can do is to describe the different perspectives and record one's disagreement with them without being able to pass any judgement on them with normative and universal force. In its most extreme form, relativism claims that **truth** is relative to intellectual backgrounds, communities, social order etc. Truth, it is said, is always 'truth for X', where X can be a person, a group, a community etc. It embeds judgement (and in particular, judgements concerning truth) in a network of norms, practices and **conventions** that may vary from one community to another. Ultimately, it equates rational judgement with **acceptance**, where the latter has no normative force or implication. There are many forms of relativism according to the domain in which the relativity is predicated. It is typically argued that relativism is self-refuting. If it is seen as a universal (absolute) claim about truth or rationality or conceptual schemes, it is obviously self-undermining. If it is seen as a relative claim, it boils down to yet another perspective that we might consider adopting, without any compelling (let alone normative) reason to adopt it.

See **Naturalism; Sociology of scientific knowledge: the strong programme**

Further reading: Baghramian (2004)

Relativity theory see **Einstein**

Reliabilism: Approach to **justification** and **knowledge** according to which a **belief** is warranted if it has been produced by a reliable process or method. It has been defended by Armstrong and Alvin Goldman (born 1938) and has been an important part of naturalised epistemology. Reliability is taken to be an objective property of a cognitive process of method, in virtue of which the belief-forming process or method generates true beliefs. On this approach, it is enough that a belief has been the product of a reliable method for it to be justified; there is no further requirement that the reliability of the process or method is independently proven or justified; nor that the believer has independent reasons to take the belief to be justified. Reliabilism shifts the focus of epistemology from the cognising subject and its transparent mind to the natural processes and methods by which knowledge can be gained and sustained. Critics of reliabilism argue that the reliability of a process or method is not enough for the **justification** of a belief because the justification of a belief has to do with what the believer does to acquire warranted beliefs, and hence with what kinds of *reasons* she demands or provides.

See **Naturalism**

Further reading: Goldman (1986)

Rules of acceptance: Rules that allow the acceptance of a conclusion in light of a set of premises. In **inductive logic,** they were also called rules of *detachment*. They were supposed to be rules that allowed the detachment of the conclusion from a set of premises, even though the conclusion followed from the premises only with (high) probability. A relatively uncontroversial example of such a rule is this: e is the total evidence; the degree of confirmation of hypothesis h in light of e is r; therefore, believe h to degree r. More interesting cases concern situations in which the degree of confirmation of a hypothesis h given the evidence e is very close to one. Would it then be reasonable to detach the probability attached to h and simply accept h? That is, would it be reasonable to move from a high *degree of belief* in h to full *belief* in it? The **lottery paradox** shows that such a rule would lead to inconsistencies. To many, **confirmation** theory (and inductive logic) does not lead to the acceptance of a hypothesis, but rather to the assignment of a **probability** to the hypothesis. Others try to entertain rules of acceptance either by denying that hypotheses have precise degrees of confirmation or by trying to avoid those cases (e.g., conjunctive beliefs) that lead to the lottery paradox.

See **Bayesianism**

Further reading: Kyburg (1974)

Russell, Bertrand (1872–1970): British philosopher with enormous impact on many areas of philosophy and the philosophy of science, one of the founders of modern mathematical logic and of analytic philosophy. His many publications include *Principia Mathematica* (with Alfred North Whitehead, 1861–1947) (1910–13), *The Problems of Philosophy* (1912), *The Analysis of Matter* (1927) and *Human Knowledge: Its Scope and Limits* (1948). His early philosophical work was characterised by its emphasis on logical analysis. He is known for his 'supreme maxim of philosophising': 'Whenever possible logical constructions are to be substituted for inferred entities'. His **principle of acquaintance** became one of the milestones of modern empiricism. He defended **empiricism**, as a thesis about the source of **knowledge**, but also accepted the subsistence of **universals** and **abstract entities** (like numbers). He defended the rationality of **induction**, arguing that the **principle of induction** is self-evident. He also aimed to reconcile empiricism with some sort of **scientific realism**, known as **structural realism**. He was initially suspicious about the concept of cause, but came to accept a number of causal principles and, in particular, an account of **causation** in terms of **causal processes** (what he called causal lines). He had a lasting commitment to **structuralism**, and though the version of structural realism advanced in *The Analysis of Matter* came under strain by a lethal objection raised by the Cambridge mathematician M. H. A. Newman (1897–1984), he kept some important structuralist commitments, for example, in his view of causation as structure-persistence.

Further reading: Psillos (1999); Russell (1927); Sainsbury (1979)

S

Salmon, Wesley (1925–2001): American philosopher of science, with ground-breaking contributions to a number of areas including the problem of **induction, causation** and **explanation**. He was the author of *Scientific Explanation and the Causal Structure of the World* (1984). He defended a mechanistic approach to causation, arguing that the missing link between cause and effect that **Hume**

was looking for is the **causal process** (**mechanism**) that connects cause and effect. He took it that the distinctive mark of causal processes is that they are capable of transmitting conserved quantities. Advancing the **statistical-relevance model of explanation**, he rejected the view that explanation are **arguments** (either deductive or inductive) and claimed that an explanation of an event consists in citing causally relevant information. He argued that the production of structure and order in the world is, at least partly, due to the existence of conjunctive forks, which are exemplified in situations in which a common cause gives rise to two or more effects. Though he favoured a causal-mechanistic account of explanation, he did argue for the importance of explanatory **unification** in science. He was an advocate of the compatibility between **empiricism** and **scientific realism** and a defender of some kind of objective **Bayesianism**, based on the idea that considerations of initial **plausibility** can ground judgements about the prior probabilities of hypotheses.

Further reading: Salmon (1984)

Scepticism: Any view that questions or doubts the possibility of **knowledge**. The typical sceptical challenge proceeds as follows. Subject S asserts that she knows that p, where p is some proposition. The sceptic asks her: how do you know that p? S replies: because I have used criterion c (or method m, or whatever). The sceptic then asks: how do you know that criterion c (or whatever) is sufficient for knowledge? It's obvious that this strategy leads to a trilemma: either infinite regress (S replies: because I have used another criterion c'), or circularity (S replies: because I have used criterion c itself), or dogmatism (S replies: because criterion c is sufficient for knowledge). One standard way to pose the sceptical challenge is via the argument from the equivalence of reasons: (1) x appears y in situation K; (2) x' appears y in situation K'; (3) we cannot discriminate between K and K' (i.e., there are no reasons to favour K over K'); hence, we cannot discriminate between x and x'. The argument from illusion is a standard example of this strategy: veridical and hallucinatory experiences appear exactly the same to the perceiving subject (they have exactly the same phenomenal content); hence, the subject cannot discriminate between perceiving a real object and hallucinating one; hence, the perceiving subject cannot have perceptual knowledge. The argument from the **underdetermination of theories by evidence** is based on similar reasoning. Scepticism can be global or local. It may concern, for instance, the very possibility of knowledge of the external world (as in **Descartes**'s case of the evil demon hypothesis) or (more locally) the possibility of knowing the existence of other minds or the existence of **unobservable entities**. There are two broad ways to address the sceptical challenge. The first is constructive: it tries to meet the challenge head-on by offering a theory of knowledge (or **justification**) that makes knowledge possible (e.g., **foundationalism**). The second is diagnostic: it denies that the sceptical challenge is natural and compelling and tries to uncover its presuppositions and to challenge them.

See **Certainty; Hume; Idealism; Induction, the problem of; Realism and anti-realism**

Further reading: Gascoigne (2002); Williams (2001)

Schlick, Moritz (1882–1936): German philosopher of science, the founder and leader of the **Vienna Circle**. He held the Chair of the Philosophy of Inductive Sciences of the University of Vienna from 1922 until his assassination by one his former doctoral students on the main staircase of the university on 22 June 1936. Before going to Vienna, he published papers on **Einstein**'s theory of relativity. His pre-positivist book *General Theory of Knowledge* was published in 1918. In this, he denied the role of intuition in **knowledge** and defended a critical version of **realism**. He developed the view that theories are formal deductive systems, where the axioms implicitly define the basic concepts. He thought, however, that implicit **definitions** divorce the theory from reality altogether: theories float freely and become a game with symbols. Given that scientific theories should have definite empirical content, Schlick argued that this content is acquired when the deductive system of the theory is applied to the empirical phenomena. In his lectures on *Form and Content* (1932), Schlick developed a structuralist understanding of science, what he called the 'geometrisation of physics' – where all content is left out leaving only pure **structure**. Throughout his career, Schlick denied the possibility of synthetic **a priori** judgements and took the view that a priori truths are analytic or conceptual truths. Schlick targeted his criticism on the so-called 'phenomenological propositions' such as 'every tone has intensity and pitch' or 'one and the same surface cannot be simultaneously red and green throughout'. He argued that these propositions are formal and analytic: they make no claim about the world; rather they assert a formal connection among concepts. Under the influence of Ludwig Wittgenstein (1889–1951) he advanced the **verifiability** criterion of meaningfulness. But he was quite clear that if realism is understood not as a metaphysical thesis but as an empirical claim, asserting that whatever is part of the spatio-temporal-causal world of science is real, it is consistent with **empiricism**. He also defended **foundationalism**.

See **Laws of nature; Protocol sentences; Structuralism**
Further reading: Ayer (1959); Schlick (1918, 1979)

Scientific method: Science is supposed to be a distinctive human enterprise and achievement partly because of its method, but there has been considerable disagreement as to what this method amounts to. There have been several candidates: **induction**, the **hypothetico-deductive method, inference to the best explanation, Mill's methods, conjectures and refutations** and others. On top of this, there has been considerable discussion concerning the **justification** of scientific method. Any attempt to characterise the abstract structure of scientific method should make the method satisfy two general and intuitively compelling desiderata: it should be ampliative *and* epistemically probative. Ampliation is necessary if the method is to deliver informative hypotheses and theories. Yet, this ampliation would be merely illusory, if the method was not epistemically probative: if, that is, it did not convey epistemic warrant to the excess content produced thus (namely, hypotheses and theories). The philosophical problem of the scientific method is whether and how these two desiderata are jointly satisfiable. Sceptics argue that they cannot be shown to be jointly satisfiable in a non-question-begging way. Popperians have tried to argue that the scientific method can refrain from being ampliative, by employing only the resources of deductive logic. Others (most notably Bayesians) have argued that a probabilistic account of the scientific method, known as **conditionalisation,** can avoid being ampliative while conferring justification on **beliefs** that have rich content. The followers of **inductive logic** have argued that the scientific method can capture an objective degree of **confirmation** of hypotheses given the evidence (via the notion of **partial entailment**). Others have aimed to face the sceptical challenge head on by trying to show how the scientific method can vindicate itself, by being self-corrective.

Advocates of **methodological naturalism** have claimed that the scientific method can be vindicated instrumentally, by reference to its past successes.

See **Ampliative inference**; **Bayesianism**; **Induction, the problem of**; **Mill**; **Naturalism**; **Objectivity**; **Peirce**

Further reading: Nola and Sankey (2000)

Scientific realism: Philosophical view about science that consists in three theses. *The Metaphysical Thesis*: the world has a definite and mind-independent structure. *The Semantic Thesis*: **scientific theories** should be taken at face value. They are truth-conditioned descriptions of their intended domain, both observable and unobservable. *The Epistemic Thesis*: mature and predictively successful scientific theories are well-confirmed and approximately true of the world. The first thesis renders scientific realism distinct from all those anti-realist accounts of science, be they traditional **idealism** and **phenomenalism** or the more modern **verificationism** of **Dummett** and **Putnam** which, based on an epistemic understanding of the concept of **truth**, allows no divergence between what there is in the world and what is issued as existing by a suitable set of epistemic practices and conditions. The second thesis – semantic realism – makes scientific realism different from eliminative **instrumentalism** and **reductive empiricism**. Opposing these two positions, scientific realism is an 'ontologically inflationary' view. Understood realistically, the theory admits of a **literal interpretation**, namely, an interpretation in which the world is (or, at least, can be) populated by **unobservable entities** and processes. The *third* thesis – epistemic optimism – is meant to distinguish scientific realism from *agnostic* or *sceptical* versions of **empiricism**. Its thrust is that science can and does deliver the truth about unobservable entities no less than it can and does deliver the truth about observable entities. It's an implicit part of the realist thesis that the ampliative-abductive methods employed by scientists to arrive at their theoretical beliefs are reliable: they tend to generate approximately true beliefs and theories.

See **Constructive empiricism; Entity realism; No-miracles argument; Pessimistic induction; Realism and anti-realism; Structural realism; Truthlikeness; Verificationism**

Further reading: Kitcher (1993); Leplin (1997); Psillos (1999)

Scientific theories see **Semantic view of theories; Syntactic view of theories**

Sellars, Wilfrid (1912–1989): American philosopher, one of the most influential and profound thinkers of the twentieth century. He is the author of *Empiricism in the Philosophy of Mind* (1956) and *Philosophy and the Scientific Image of Man* (1960). He defended a reformed version of **empiricism**, freed from **foundationalism**. In his attack on the **given**, he argued that experience bears on theories not by providing an incorrigible and theory-free foundation, but by putting theories in jeopardy. He also distanced himself from **coherentism**. Empirical knowledge, for Sellars, rests on the self-corrective **scientific method**. Sellars was a fierce critic of **instrumentalism**. His defence of **scientific realism** was based on the claim that science gives the ultimate explanation of what the world is like and that this explanation is complete and adequate only by reference to **unobservable entities** and their **properties**. He argued that scientific **explanation** proceeds via the *theoretical identifications* of observable entities with their unobservable constituents. In his prioritising the scientific image of the world over the manifest image, Sellars took it that science is the measure of all things. He warned us against confusing the right idea that philosophy is not science with the mistaken idea that philosophy is independent of science.

Further reading: deVries (2005); Sellars (1963)

Semantic realism see **Feigl; Scientific realism**

Semantic view of theories: Cluster of views about theories that place **models** at centre-stage. The core of this view is that theories represent the world by means of models and hence that the characterisation of theories, as well as the understanding of how they represent the world, should rely on the notion of model. Where the logical positivists favoured a formal axiomatisation of theories in terms of first-order logic, thinking that models can play only an illustrative role, the advocates of the semantic view opted for a looser account of theories, based on mathematics rather than meta-mathematics. To be sure, a strand within the semantic view, followed primarily by Sneed and the German structuralists (e.g., **Stegmuller**) aimed at a formal set-theoretic explication (and axiomatisation) of scientific theories. But the general trend was to view theories as clusters of (mathematical) models. As an answer to the question 'What is a scientific theory?', the semantic view claims that a scientific theory should be thought of as something *extra-linguistic*: a certain **structure** (or class of structures) that admits of different linguistic garbs. This thought goes back to **Suppes** and was further pursued by Fred Suppe (born 1940) and **van Fraassen**. A key argument for the semantic view is that it tallies better with the actual scientific conception of theories. For instance, it is more suitable for biological theories, where there is no overarching axiomatic scheme. Besides, it does not fall prey to the problems that plagued the **syntactic view of theories**. A challenge to this view is that it is unclear how theories can represent anything empirical and hence how they can have empirical content. This challenge has been met in several ways, but two ways have been prominent. The first way is that the representational relation is, ultimately, some mathematical *morphism*. The theory represents the world by having one of its models isomorphic to the world or by having the empirical phenomena embedded in a model of the theory. However, mathematical morphisms preserve only structure and hence it is not clear how a theory acquires any specific empirical content, and in particular how it can be judged as true. The second way is that theories should be seen as mixed entities: they consist of mathematical models plus theoretical hypotheses. The latter are linguistic constructions which claim that a certain model of the theory represents (say by being similar to) a certain worldly system. Theoretical hypotheses have the form: the physical system *X* is, or is very close to, *M* – where *M* is the abstract entity described by the model. On this view, advanced by **Giere** and endorsed by van Fraassen, theoretical hypotheses provide the *link* between the model and the world.

Further reading: Giere (1988); Suppe (1989); van Frasssen (1980)

Sense and reference: Central concepts in the theory of meaning. According to **Frege**'s early theory of meaning, the reference (semantic value) of an expression is that feature of the expression that determines its contribution to the truth (or falsity) of the sentences in which it occurs. In particular, the reference of a proper name is the object it refers to or stands for, the extension of a predicate is the class of things to which it applies, and the semantic value of a sentence is its truth-value (truth or falsity). Later on, Frege introduced *senses* in his theory of meaning in order to explain a difference in the knowledge of identity statements such as 'The morning star is the morning star' and 'The Morning Star is the Evening Star'. Though both expressions ('morning star' and 'evening star') have the same reference, namely, the planet Venus, the first statement is trivial and can be known **a priori** while the second is informative and can be known a posteriori. Frege explained this difference by arguing that the two expressions have the same reference (semantic value), but differ in their senses. These are modes of presentation of what is designated. Frege took the sense of an expression to be what someone who understands the expression grasps. He took the sense of a sentence to be a thought. Subscribing to anti-psychologism, Frege took senses (and hence thoughts) to be objective – they specify a condition such that when it obtains it is necessary and sufficient for the **truth** of the sentence that expresses the thought. A complete theory of meaning then should be two-dimensional: it should take the meaning of an expression to include both its sense and its reference. The Fregean orthodoxy was challenged by **Kripke**.

See **Causal theory of reference**

Further reading: Devitt and Sterelny (1987)

Sense data: The content of experience – what a subject senses. They have been posited to account, among other things, for the phenomenological (or qualitative) similarity between veridical and hallucinatory experiences. A sense datum is supposed to be the common factor between them. (As, for instance, when I see a green leaf and when I am hallucinating a green leaf – the green-like image that is common to both experiences is a sense datum.) Accordingly, sense data are supposed to be mental items, though some philosophers have taken them to be neutral elements. For some empiricists, they constitute the incorrigible foundation of **knowledge**, though this view has been criticised by **Sellars** in his attack on the myth of the **given**. If sense data are taken seriously, the issue crops up of how they are related, if at all, with material objects. Phenomenalism is the view that material objects are constituted by actual and possible sense data. Some versions of it claim that talk of material objects is fully translated into talk about sense data. But this last view has been discredited, partly because this translation would require the **truth** of certain **counterfactual conditionals** (e.g., if I were to look at so-and-so, I would experience such-and-such sense data), and it is hard to see what other than material objects could be the truth-makers of these conditionals.

See **Empiricism; Foundationalism**

Further reading: Huemer (2001)

Simplicity: Virtue of **scientific theories**. Though most philosophers of science think that simpler theories are to be preferred to more complicated ones, there is little agreement as to (1) how simplicity is to be understood and (2) what the status of this virtue is. As to (2) it is argued that simplicity is a pragmatic or aesthetic virtue such that its possession by a theory does not affect its **probability** to be true. Those philosophers of science who take simplicity to be a cognitive or epistemic virtue face the problem of justifying why simpler theories are more likely to be true than more complicated ones. A possible **justification** would be related to the view that the world itself is simple; but this would be a metaphysical commitment that would be difficult to justify **a priori**. An a posteriori justification – based on the empirical successes of simple theories – would be more adequate, but it would still require commitment to the claim that inductive reasoning is reliable. As to (1) above, if simplicity is understood as a syntactic property, it will vary with the formulation of the theory. Yet, there seems to be a strong connection between simplicity and adjustable parameters: the more adjustable parameters a theory has, the more complex it is. If simplicity is understood in a more ontological way, it should be connected with the number of entities posited by the theory. Here again, there is a difference between *types* of entity and tokens of entities. Normally, simplicity is connected to the number of types of entity that are posited.

See **Ad hocness/ ad hoc hypotheses; Curve-fitting problem; Ockham's razor**

Further reading: Swinburne (1997)

Smart, J. J. C. (born 1920): British-born Australian philosopher of science, author of *Philosophy and Scientific Realism* (1963). He has been one of the first and most vigorous defenders of **scientific realism** and the author of a version of the **no-miracles argument**. Smart argued against **instrumentalism** that it implies the existence of a cosmic coincidence. He was also a key figure in the advancement of materialism; he defended the view that mental states are identical with physical states (an identity which he took it to be theoretical and a posteriori) arguing that it gives the simpler and most comprehensive **explanation** of the working of the mind that is consistent with the empirical findings of the brain sciences.

Further reading: Smart (1963)

Social constructivism: Agglomeration of views with varying degrees of radicalness and plausibility. Here is a sketchy list of them. The acceptability of a **belief** has nothing to do with its **truth;** beliefs are determined by social, political and ideological forces, which constitute their causes. Scientific facts are constructed out of social interactions and negotiations. Scientific objects are created in the laboratory. The acceptability of **scientific theories** is largely, if not solely, a matter of social negotiation and a function of the prevailing social and political values. Science is only one of any number of possible 'discourses', none of which is fundamentally truer than any other. What unites this cluster of views are (vague) slogans such as 'scientific truth is a matter of social authority' or 'nature plays (little or) no role in how science works'. It might be useful to draw a distinction between a weak form of social constructivism and a strong one. The weak view has it that some categories (or entities) are socially constructed: they exist because we brought them into existence and persist as long as we make them so. Money, the Red Cross and football games are cases in point. But this view, though not free of problems, is almost harmless. On the strong view, all reality (including the physical world) is socially constructed: it is a mere projection of our socially inculcated conceptualisations.

See **Relativism; Sociology of scientific knowledge: the strong programme**

Further reading: Nola (2003)

Sociology of scientific knowledge: the strong programme Programme for doing sociology of science advanced by Barry Barnes (born 1943) and David Bloor (born 1942), who have founded the so-called Edinburgh School. It is contrasted to what they called the 'weak program' for doing sociology of science, according to which sociological explanations of scientific beliefs are legitimate but only in so far as scientific beliefs are irrational or otherwise inappropriate (e.g., biased, mistaken etc.). According to the strong programme, all beliefs should be subjected to the same type of causal **explanation**. The four tenets of the strong programme are:

Causality: the explanation of **belief** should be causal, that is, it should be concerned with the conditions (psychological, social and cultural) that bring about beliefs.

Impartiality: the explanation of belief should be impartial with respect to the traditional dichotomies such as **truth**/falsity, **rationality**/irrationality and success/failure in that both sides of these dichotomies require explanation.

Symmetry: the same types of cause should be used in the explanations of true and false beliefs.

Reflexivity: the explanation of belief should be reflexive in that the very same patterns of explanation should be applicable to sociology itself.

Though very influential among sociologists of science, the strong programme has been accused for leading to **relativism**.

Further reading: Bloor (1991); Koertge (1998); Nola (2003)

Space: God's sensorium, according to **Newton**, who thought that space was absolute: independently existing, unchangeable and immovable; the state of absolute rest. **Leibniz** argued against Newton that space is nothing over and above the spatial relations among material objects; hence it is an **abstraction**. **Kant** argued that space (and **time**) is an **a priori** form of pure intuition; the subjective condition of sensibility, without which no experience would be possible. Hence, he took it that space does not represent any properties of things-in-themselves; space, as he put it, is empirically real and transcendentally ideal. He identified the form of (outer) intuition with **Euclidean geometry**, thinking that he could thereby secure the knowledge of the phenomenal world as this is expressed by Newtonian mechanics. The advent of **non-Euclidean geometries** challenged this Kantian view and **Hilbert's** axiomatisation of geometry removed the privileged status of spatial intuition. **Einstein**'s theory of relativity made use of non-Euclidean geometries and at the same time denied that there is such a thing as absolute space. Einstein's relativity merged space and time into a four-dimensional manifold: **spacetime**.

See **Grünbaum; Poincaré; Reichenbach**

Further reading: Earman (1989)

Spacetime: According to the Russian-German mathematician Hermann Minkowski (1864–1909), the four-dimensional continuum with three spatial dimensions and one temporal one (known as spacetime) is more fundamental than space or time. As he claimed, **space** by itself and **time** by itself fade away into shadows, their union (spacetime) being the only reality. He presented **Einstein**'s special theory of Relativity within this four-dimensional metric structure. Einstein went on to adopt and develop this notion. There have been two broad views about the nature of spacetime. According to substantivalism, spacetime is some sort of substance. According to relationism, spacetime is the collection of the spatio-temporal relations among material objects.

Further reading: Sklar (1974)

Statistical-relevance model of explanation: Account of **explanation** developed by **Salmon** in an attempt to improve on the **inductive-statistical model**. In judging whether a further factor C is relevant to the explanation of an event that falls under type E, we look at how taking C into account affects the **probability** of E to happen. In particular, a factor C explains the occurrence of an event E, if $\text{prob}(E/C) > \text{prob}(E)$ – which is equivalent to $\text{prob}(E/C) > \text{prob}(E/not\text{-}C)$. Note that the actual values of these probabilities do not matter. Nor is it required that the probability prob(E/C) be high. All that is required is that there is a *difference*, no matter how small, between the two probabilities.

See **Causation**

Further reading: Psillos (2002); Salmon, Jeffrey and Greeno (1971)

Statistical testing: Testing of statistical hypotheses. In its classical formulation, it is a hybrid of the approaches of R. A. Fisher (1890–1962) and Jerzy Neyman (1894–1981) and Egon Sharpe Pearson (1895–1980). In *The Design of Experiments* (1935), Fischer introduced the idea of a null hypothesis, and argued that statistical **inference** is concerned with the rejection of the null hypothesis. This is achieved when the sample estimate deviates from the mean of the sampling distribution by more than a specified percentage, the level of significance – which Fisher took it to be 5 per cent. Fischer was an advocate of **falsificationism**, arguing that experiments exist 'in order to give the facts a chance of disproving the null hypothesis'. Neyman and Pearson's statistical methodology was originally conceived as an attempt to improve on Fisher's approach. This method formulates two hypotheses, the null hypothesis and the alternative hypothesis – capitalizing on the methodological rule that hypotheses should be tested against alternatives as well as on the rule that a hypothesis is not rejected unless there is another one to replace it. In the Neyman–Pearson framework, there are two types of error: a *true* null hypothesis can be incorrectly rejected (Type I error) and a *false* null hypothesis may fail to be rejected (Type II error). Neyman and Pearson thought that avoiding Type I error is more important than avoiding Type II error – hence the design of experiment should be such that the test should reject the hypothesis tested when it is true very infrequently. As they said, the issue is not so much whether a single hypothesis is true or false, but rather devising a rule for testing hypotheses such that 'in the long run of experience, we shall not be too often wrong'.

Further reading: Hacking (1965); Mayo (1996)

Stegmuller, Wolfgang (1923–1991): German philosopher of science, author of *The Structuralist View of Theories* (1979). He was one of the chief advocates of set-theoretic **structuralism**, the view that a theory is identified with a set-theoretic predicate. The structure of the theory is then presented in terms of the connections between the **models** of the theory, the intended applications of the theory etc. His followers, notably C. Ulises Moulines (born 1946) and Wolfgang Balzer (born 1947), have further developed **structuralism** by applying it to traditional philosophical problems such as the structure of **explanation** or the inter-theoretic **reduction**. This structuralist approach was initiated by Joseph Sneed (born 1938) in *The Logical Structure of Mathematical Physics* (1971).

See **Semantic view of theories**

Further reading: Stegmuller (1979)

Straight rule of induction: Rule of inductive **inference** advocated by **Reichenbach**. Given that the actual relative frequency of observed *A*s that are *B*s is m/n, we should have m/n as our degree of confidence regarding the *B*-hood of the next *A*. If, in particular, all observed *A*s have been *B*s (i.e., if m = n), then the rule tells us that we should assign **probability** one that the next *A* will be *B*.

See **Induction, the problem of; Laplace**

Further reading: Salmon (1967)

Structural realism: Philosophical position concerning what there is in the world and what can be known of it. In its stronger form, it is an ontic position: **structure** is all that there is. In its weaker form, it is an epistemic position: there is more to the world than structure, but only the structure of the world can be known. The epistemic position has had two kinds of input. The *first* goes back to **Russell** who advanced a structuralist account of our knowledge of the world, arguing that only the structure, that is, the totality of formal, logico-mathematical properties, of the external world can be known, while all of its **intrinsic** (qualitative) **properties** are inherently unknown. Russell claimed that the logico-mathematical structure can be legitimately *inferred* from the structure of the perceived phenomena. The second input goes back to the writings of **Poincaré** and **Duhem**. Their structuralism was motivated by the perceived discontinuities in theory-change in the history of science (also known as scientific revolutions) and aimed to show that there is continuity at the level of the structural description of the world: the structure of the world could be revealed by structurally-convergent scientific theories. These two inputs have been united into what came to be known as structural realism in the writings of Grover **Maxwell** and, in the 1980s, of John Worrall (born 1946) and **Zahar**. The twist they gave to structuralism was based on the idea of **Ramsey-sentences**. Given that the Ramsey-sentence captures the logico-mathematical form of the original theory, the structuralist thought is that, if true, the Ramsey-sentence also captures the structure of reality: the logico-mathematical form of an empirically adequate Ramsey-sentence mirrors the structure of reality. Yet it turns out that unless some non-structural restrictions are imposed on the kinds of things the existence of which the Ramsey-sentence asserts, that is, unless structuralism gives up on the claim that *only* structure can be known, an empirically adequate Ramsey-sentence is bound to be true: **truth** collapses to **empirical adequacy**. The ontic version of structural realism, defended by James Ladyman (born 1969) and Steven French (born 1956), aims to eliminate objects altogether and takes structures to be ontologically primitive and self-subsistent.

See **Entity realism; Scientific realism**

Further reading: Da Costa and French (2003); Ladyman (2002); Psillos (1999); Worrall (1989)

Structuralism: Cluster of views that prioritise **structure** over content. Typically, structure is viewed as a system of relations, or as a set of equations. The content of a structure is then viewed as the entities that instantiate a structure. Extreme forms of structuralism take the structure in a purely formal way, leaving to one side the interpretation of the relations and focusing only on their formal (logical-mathematical) properties. Structuralism in the philosophy of science comes in many brands and varieties. It ranges from a methodological thesis (associated with the **semantic view of theories**) to a radical ontic position (claiming that structure is all there is). In between, there is an epistemic view: there is more to the world than structure, but of the world nothing but its structure can be known.

See **Maxwell, Grover; Poincaré; Structural realism**

Further reading: Da Costa and French (2003)

Structure: A *relational system*, a collection of objects with certain **properties** and relations. A structure is the *abstract form* of this system. Focusing on structure allows us to take away all features of the objects of the system that do not affect the way they relate to one another. More formally, two classes of objects A and B have the same structure (they are *isomorphic*) iff there is an one–one correspondence f between the members of A and B and whenever any n-tuple $<a_1 \ldots a_n>$ of members of A stand to relation P their image $<f(a_1) \ldots f(a_n)>$ in B stands to relation $f(P)$, where $f(P)$ is the image of P in B.

Further reading: Da Costa and French (2003)

Supervenience: Modal relation of determination that exists between two domains, or two sets of **properties**, or two sets of facts A and B. In slogan form: A supervenes on B if there is no A-difference without a B-difference – for example, no mental difference without a physical difference; no aesthetic difference without a physical difference. Properties A and B (e.g., mental properties and neurophysical properties) might well be distinct and separate, and yet it might be that the A properties supervene on the B properties in that two entities X and Y that are alike in all B properties are also alike in their A properties. The metaphysical importance of this relation is that it accounts for the ontic priority of some facts or properties (the so-called subvenient basis) without denying *some* kind of autonomy to the supervenient facts or properties.

See **Humean supervenience; Reduction**
Further reading: Kim (1993)

Suppes, Patrick (born 1924): American philosopher of science and logician, author of *Probabilistic Metaphysics* (1984). He has worked on the philosophy of **causation**, defending a probabilistic account. He has also done pioneering work on **models** and has been one of the founders of the **semantic view of theories**. Suppes has taken indeterminacy and uncertainty seriously and has tried to develop a philosophical theory of science that does justice to both.

Further reading: Suppes (1984)

Symmetry thesis see Sociology of scientific knowledge: the strong programme

Syntactic view of theories: Also known as the received view, it identifies theories with (pieces of) languages. In its strong version, the received view took it that the language of first-order logic provided the framework in which the syntactic structure of a theory could be cast. As developed by **Carnap**, it brought together the **Duhem-Poincaré** view that theories are systems of hypotheses whose ultimate aim is to save the phenomena, and the **Hilbert** formalisation programme, according to which theories should be reconstructed as formal axiomatic systems. Many empiricists thought that a scientific theory need not be fully interpreted to be meaningful and applicable. They thought it enough that only *some* terms, the so-called observational terms and predicates, be interpreted. What confers partial interpretation on theoretical terms is a set of **correspondence rules** that link them with observational terms. But, in their attempt to put meaning back into the formal structure of the theory, empiricists got themselves into all sorts of problems over the issue of the meaning of theoretical terms. By identifying theories with formal languages the syntactic view drastically impoverished theories as means of representation. It is often more practical – and even theoretically more plausible – to start with a class of **models** and then inquire whether there is a set of axioms such that the models in the given class are its models.

See **Holism, semantic; Semantic view of theories; Terms, observational and theoretical**

Further reading: Carnap (1956); Suppe (1977)

Synthetic a priori see **A priori/a posteriori; Kant**

T

Tacking paradox, the: A problem faced by many theories of confirmation. Take a hypothesis *H* which entails some piece of evidence *e*. Add to *H* any irrelevant statement *H'* whatever (e.g., that God exists or that the absolute is sleepy or what have you). Since the conjunction *H* and *H'* also entails *e*, it is confirmed by *e* too. Hence *H'* (a totally irrelevant statement) is also confirmed by the evidence. This problem is particularly acute for the **hypothetico-deductive** method of confirmation, but it also plagued **Hempel's theory of confirmation** (via the converse consequence condition) and it affects the Bayesian account as well. It is also called the irrelevant conjunction problem or the problem of isolated statements. **Carnap** tried to solve it by claiming that a theoretical statement is meaningful and confirmable *not* just in case it is part of a theory, but rather when it makes some positive contribution to the experiential output of the theory.

Further reading: Hacking (1965)

Teleology see **Functional explanation**

Terms, observational and theoretical: Terms and predicates like 'table', 'pointer', 'is red', 'is square', 'is heavier than' have been called observational because they are supposed to get their meaning directly from experience: the conditions under which assertions involving them are verified in experience coincide with the conditions under which they are true. They were contrasted to theoretical terms that were supposed to get their meaning via theory. Many empiricists took theoretical terms to be semantically suspect and got involved in a number of projects aiming to account for their meaning on the basis of the meaning of observational ones. The very distinction between these two types of terms was challenged in the 1960s when the thesis that all observation is theory-laden became popular as many philosophers espoused semantic **holism**.

See **Correspondence rules; Definition, explicit; Observation, theory-ladenness of**

Further reading: Carnap (1956); Psillos (1999)

Theoretical terms see **Terms, observational and theoretical**

Theoretical virtues: Properties, such as **simplicity**, fertility, naturalness, unity, lack of **ad hoc** features, that characterise a good theory. They are called virtues precisely because a theory that possesses them is considered virtuous. Occasionally, explanatory power is considered an independent theoretical virtue, but some philosophers take it that the explanatory power of a theory is constituted by virtues such as the above. **McMullin** has drawn a useful distinction between synchronic virtues (such as logical consistency or simplicity) and diachronic virtues (such as fertility and **consilience of inductions**) that characterise the development (and the potential) of a theory over time. Diachronic virtues are epistemically significant because they relate to how a theory responds to pressure that comes from the **evidence,** or from other theories. A theory that yields **novel predictions,** for instance, is more credible than a theory that gets modified in an ad hoc way in order to fit with the data. Other philosophers, however, take all virtues to be pragmatic or aesthetic.

Further reading: McMullin (1992)

Theoretician's dilemma: It was introduced by **Hempel** and was related to **Craig's Theorem.** If the theoretical **terms** and principles of a theory do not serve their purpose of a deductive systematisation of the empirical consequences of a theory, they are dispensable (unnecessary). But, given Craig's theorem, even if they do serve their purpose, they can be dispensed with since any theoretical assertion establishing a connection between observational antecedents and observational consequents can be replaced by an observational assertion that directly links observational antecedents to observational consequents. Theoretical terms and principles of a theory either serve their purpose or they do not. Hence, the theoretical terms and principles of any theory are dispensable. One of the problems this argument faces is that it rests on an implausible distinction between theoretical statements and observational ones. Another problem is that it patently fails to account for the indispensable role of theories in establishing **inductive systematisations** of observational statements.

See **Instrumentalism**

Further reading: Hempel (1965)

Thought experiment: Way of testing a hypothesis by imagining or thinking what would happen (what could be observed; what difference would follow) if this hypothesis were true. Whether or not this is an experiment is controversial. But this technique has been quite popular among philosophers and many scientists, including the likes of **Galileo**, **Newton** and **Einstein**. The expression 'thought experiment' (*Gedankenexperiment*) was popularised (though not invented) by **Mach**. Einstein, for instance, argued for his principle of equivalence by imagining an observer confined within a lift and by claiming that for him there is no way to distinguish between being at rest in the earth's gravitational field and being accelerated by a rocket. And Newton argued for **space** being absolute based on his rotating bucket thought experiment. Pretty much as in ordinary experiments, a thought experiment needs to consider alternative hypotheses and explanations, as well as to take account of idealisations and **abstractions**.

Further reading: Brown (1991)

Time: In his *Confessions*, St Augustine (354–430) famously said that he knew what time was so long as no one asked him to explain it. Then, he went on to argue that time flows: the past *has been*, the future *will be* and the present *is*. **Kant** argued that time (as well as **space**) is an **a priori** form of pure intuition. **Newton** took it to be absolute: not only is there an absolute fact of the matter as to what events are simultaneous but also time is the substratum (template) within which all physical **events** are embedded and occur. **Leibniz** advanced a relational account of time: time is fully determined by the relations that exist among events. **Einstein** argued that the relation of simultaneity (as well as duration) is frame-dependent. Time seems to have a preferred direction: it is asymmetric. This *arrow of time* is puzzling, since the laws of fundamental physics are time-symmetric. But there are macroscopic processes that are irreversible. Does then the direction of time emerge at the macroscopic level? Many think that the arrow of time is thermodynamic: it is sustained by the second law of thermodynamics according to which, in a closed system, entropy increases. Others take the line that time's arrow is simply the causal arrow: that causes precede their effect. Does time flow? There have been two theories concerning this, that go back to J. M. E. McTaggard (1866–1925). On the a-series approach, events are ordered in time according to their being in the past, present and future. This is known as the moving-now theory: the now is like a luminous dot moving along a straight line; the real is whatever is illuminated by the dot; everything else is unreal in that it has either ceased to exist or has not existed yet. On the b-series approach, events are ordered in time according to the earlier-than relation. There is, then, no privileged 'now', nor any kind of flow of it. All events are equally real – since they have a definite place in the series. McTaggard argued for the unreality of time. Many philosophers have taken the view, known as 'the block universe theory' or 'eternalism', that there are no significant ontological differences among present, past and future. The apparent difference between, say, *now* and *past* is explained by trying to reduce a temporal indexical proposition (e.g., I am having a terrible toothache *now*, or I had a terrible toothache *yesterday*) to some temporal non-indexical proposition that claims that either two events are simultaneous or that they stand to the earlier-than relation.

See **Spacetime**

Further reading: Le Poidevin and MacBeath (1993)

Total evidence, principle of: Methodological principle of **inductive logic**: in determining the degree of **confirmation** of a hypothesis in the light of the **evidence**, one should rely on the *total* (observational) evidence available. The application of this principle is necessary in inductive logic because the degree of confirmation of a hypothesis in the light of some evidence can be substantially reduced if further evidence is taken into account.

See **Inductive-statistical model of explanation**

Further reading: Carnap (1950b); Skyrms (2000)

Tropes: Particularised **properties**. Taken as tropes, properties are particulars that exist independently of each other and combine to constitute the several and varied entities that make up the world. Traditional **particulars** (individuals) are said to be collections of compresent tropes. Traditional **universals** are equivalence classes of perfectly resembling tropes. The notion of resemblance is left unanalysed; yet, it is argued that it admits of degrees: tropes resemble each other more or less. The notion of compresence is also left unanalysed; yet, it is argued that compresence relates to co-location. Here is an example. Two white books sit on the desk. There are two concrete particulars on the desk. But (1) we don't have, in this situation, two instances of the universal *whiteness*, which is wholly present in the two books. Instead, two distinct (but resembling) abstract particular whitnesses are at two distinct locations on the desk. And (2) the two concrete particulars (i.e., the two books) are what they are (and distinct from each other) in virtue of the different compresent collections of tropes that 'make them up'. Trope theory is attractive on many counts, but mostly because it is ontically parsimonious. It uses just one type of building block of the **reality**. According to the memorable expression of the Harvard professor D. C. Williams (1899–1983) tropes are 'the very alphabet of being'.

Further reading: Campbell (1990)

Truth: There are two strands in our thinking about truth. The *first* is to say that truth is an *objective* property of our beliefs in virtue of which they correspond to the world. Truth connects our thoughts and beliefs to some external reality, thereby giving them representational content. Truth is then an external constraint on what we believe. The second strand takes it that truth is an *evaluative* and *normative* concept: it summarises the norms of correct assertion or **belief**: to say of a belief that it is true is to say that it is epistemically right, or justified, to have it. Hence, truth is an internal constraint on what we believe. Whatever else it is, truth does not have an expiry date. Unlike dairy products, truth cannot go off. Hence, truth cannot be equated with acceptance. Nor can it be equated with what communities or individuals agree on, or with what the present evidence licences. If we made these equations, truth would not be a stable property of beliefs. It is crucial, when we think of truth as an evaluative concept, to think of the norms that govern its use as objective. One way to develop this view is the coherence theory of truth, according to which a belief is true if and only if it is a member of a coherent system of beliefs. Apart from problems that have to do with how exactly the notion of coherence should be understood, this way of developing the evaluative approach has met with the further difficulty that it cannot satisfy Tarski's definition of truth – which has been taken to provide an adequacy constraint on any theory of truth. A much more promising way to develop the evaluative approach has been advanced by **Dummett** and his followers. This is based on a justificationist understanding of epistemic rightness: it has equated truth with warranted assertibility. The truth of an assertion is conceptually linked with the possibility of recognising this truth. Recently, Crispin Wright (born 1942) has strengthened this account by taking truth to be superassertibility, this being a kind of strong assertibility which would be endurable under any possible improvement to one's state of information. Both the objective and the evaluative strands take truth to be a substantive property of truth-bearers: to say of a belief that it is true is to attribute a substantive property to it. But they disagree over the nature of this property. The objective account takes this property to be non-epistemic, namely, a property that a belief has independently of any knowledge of it that the knowing subject might have. The evaluative account takes truth to be an epistemic property, namely, a property that beliefs have because and in so far as they can be known to be true (e.g., they can be verified, or proved etc.). The difference between a non-epistemic and an epistemic conception of truth becomes evident if we think in terms of the Socratic Euthyphro contrast: are statements true *because* they are licensed as true by a set of norms or are they licensed as true by a set of norms *because* they are true? A moment's reflection will show that there is all the difference in the world in taking one or the other side of the contrast.

See **Truth, semantic conception of**

Further reading: Kirkham (1992); Vision (2004); Wright (1992)

Truth, coherence theories of see **Truth**

Truth, correspondence theories of see **Truth**

Truth, deflationary approach to: Family of views that focus on the role of the truth-predicate in language and ascribe a quasi-logical, or expressive, function to it, that it is useful for forming generalizations of a particular kind. They are said to deflate *truth* because they deny that the truth-predicate stands for a substantive or complex (epistemic or non-epistemic) property, in particular a property that can play an explanatory role. The logical need that the truth-predicate is supposed to cover is captured by statements of the form 'Whatever Plato said was true': instead of saying Plato said that p and p and Plato said that q and q etc., we form the foregoing generalisation. Deflationists typically argue that Tarski's convention T (an instance of which is the famous sentence 'Snow is white' is true if and only if snow is white) captures all there is to truth and offers an implicit **definition** of the truth-predicate. Convention T has been taken to be a disquotational schema: it provides means to remove the quotation marks around the name of a sentence (to say that a sentence is true is to assert this sentence). Hence, it has been taken to offer a disquotationalist account of **truth**.

Other deflationists (notably **Ramsey**) have favoured the so-called redundancy theory of truth, according to which all there is to truth is captured by the schema: it is true *that p iff p*, where *p* is a variable ranging over propositions (expressed by sentences of a language *L*). Though advocates of deflationism have claimed that their account is explanatorily complete (it explains everything there is to know about the role of a truth-predicate in a language), critics of deflationism claim that there are salient facts about truth that are not explained by the deflationary approach. One particularly acute problem concerns the cases where translation from one language into another is involved.

See **Truth, semantic theory of**

Further reading: Horwich (1998b); Wright (1992)

Truth, pragmatic theory of see **James; Peirce; Pragmatism; Truth**

Truth, semantic conception of: Alfred Tarski's (1902–1983) conception of **truth**: truth expresses relations between linguistic entities and extra-linguistic structures or domains. Tarski suggested that the T-sentence of the form ' "Snow is white" is true if and only if snow is white' captures what it is for a sentence to be true. He required that the truth-predicate for a language *L* be introduced in a meta-language (in order that formal paradoxes such as the Liar **paradox** are avoided) and that it must be such that it is materially adequate: the definition of the truth-predicate must yield all T-sentences of the object language *L*. These T-sentences are instances of the meta-linguistic schema *T* (the convention *T*): *X* is true-in-*L* if and only if *p*, where *X* stands for the (meta-linguistic) names of the sentences of *L*; 'is true-in-*L*' is the truth-predicate defined in the meta-*L*; and *p* stands for the translations in the meta-*L* of the corresponding sentences of L. Tarski advanced a *recursive* introduction of the truth-predicate by means of operations on atomic formulas – that is, sentential functions which, in their simplest form, contain a monadic predicate followed by a (free) variable, namely, Fx. Sentential functions are neither true nor false. In order for these categories to apply, the sentential functions must be replaced by sentences, that is, their free variables must get values. If, for instance, the variable x in the sentential function Fx gets object **a** as its value, one can say that the resulting sentence Fa is true if and only if **a** satisfies (belongs to the extension of) F and false otherwise. In other words, Tarski introduced the truth-predicate by means of the notion of *satisfaction*, which, according to him, can be rigorously defined. The notion of satisfaction is akin to that of reference. It has been argued that Tarski's technique gives only a definition of 'true-in-L' and not of 'true for variable L'. Other philosophers claim that, supplemented with a **causal theory of reference**, Tarski's account of truth is substantive and captures the idea that truth is correspondence with reality.

See **Sense and reference; Truth, deflationary approach to**

Further reading: Tarski (1944, 1969)

Truth-maker principle: It states that for every contingent **truth** there must be something in the world that *makes* it true. Truth-makers of propositions are said to be states of affairs (e.g., an object having a certain property or two or more objects standing in a certain relation). Objects can also be truth-makers, for example, that planet Mars is the truth-maker of the proposition that Mars exists.

See **Universals**

Further reading: Armstrong (2004)

Truthlikeness: Concept introduced by Graham Oddie (born 1954) and Ilkka Niiniluoto (born 1946) in an attempt to correct the flaws of **Popper**'s definition of **verisimilitude**. It is meant to capture the distance between a possible world and the actual world. The actual world is one of the possible worlds. A theory T is *true* if and only if it describes the actual world. A false theory may, nonetheless, be truthlike in that the possible world it describes may agree on some facts with the actual world (described by the true theory). This partial agreement is employed to explicate formally the notion of truthlikeness. But formal theories of truhlikeness face significant problems. Chief among them is that the degree in which a certain theory is truthlike will depend on the language in which the theory is expressed. In particular, two logically equivalent theories may turn out to have different degrees of truthlikeness.

See **Pessimistic induction**
Further reading: Oddie (1986)

Underdetermination of theories by evidence: Evidence is said to underdetermine theory. This may mean two things. *First*, the **evidence** cannot prove the **truth** of the theory. *Second*, the evidence cannot render the theory probable. Both kinds of claim are supposed to have a certain epistemic implication, namely, that belief in theory is never warranted by the evidence. *Deductive underdetermination* rests on the claim that the link between evidence and (interesting) theory is *not* deductive. But this does not create a genuine epistemic problem. There are enough reasons available for the claim that belief in theory can be justified even if the theory is not proven by the evidence: warrant-conferring methods need not be deductive. *Inductive underdetermination* rests on two major arguments that question the confirmatory role of the evidence vis-à-vis the theory. The first capitalises on the fact that no evidence can affect the **probability** of the theory unless the theory is assigned some non-zero prior probability. The second rests on the claim that theories that purport to refer to unobservable entities are, somehow, unconfirmable. It is sometimes argued that for any theory we can think of there will be totally empirically equivalent rivals, that is, theories that entail exactly the same observational consequences under any circumstances. This empirical equivalence thesis is an entry point for the epistemic thesis of total underdetermination, namely, that there can be no evidential reason to believe in the truth of *any* theory. But there is no proof of the empirical equivalence thesis, though a number of cases have been suggested ranging from **Descartes**'s 'evil demon' hypothesis to the hypothesis that for every theory *T* there is an empirically equivalent rival asserting that *T* is empirically adequate-yet-false, or that the world is *as if T* were true. One can argue that these rival hypotheses have only philosophical value and drive only an abstract philosophical **scepticism**.

See **Duhem–Quine thesis**

Further reading: Laudan (1996)

Unification: A central aim of intellectual inquiry. It has been standardly assumed that the aim of science is understanding the world by systematising all facts into a unified theoretical system. A long instrumentalist tradition, going back to **Mach** and **Duhem**, has taken unification to be an independent aim of science over and above the aim of saving the phenomena. Mach tied unification to his view of science as economy of thought, while Duhem argued that unification amounts to a natural classification of the phenomena. However, they both thought that a unified theoretical system need not be true of the world. More realistically minded philosophers of science have taken it to be the case that the world itself is unified, but recently **Cartwright** and others have argued that the world is disunified. Michael Friedman (born 1947) has argued that if a number of seemingly independent regularities are shown to be subsumable under a more comprehensive law, our understanding of nature is promoted: the number of regularities which have to be assumed as 'brute' is minimised. An alternative approach has been developed by Philip Kitcher (born 1947) who takes it that unification is achieved by minimising the number of explanatory patterns or schemata.

See **Explanation, unification model of; Laws of nature**
Further reading: Morrison (2000)

Unity of science: Doctrine favoured by the logical positivists in the 1930s and 1940s. They advocated the unity of science as an **a priori** principle that aimed to bring all scientific concepts under one-and-the-same framework. Physics was taken to be the fundamental science, on the basis of which all other scientific concepts should be defined. The unity of science was conceived of primarily as a linguistic doctrine: the unity of the language of science. The logical positivists took this unity of language for granted because they thought that (1) new terms (or concepts) should be admitted only if there is a method that determines their meaning by reference to observations and (2) this method of determination relates, ultimately, to the intersubjective language of physics. Accordingly, if the concepts of the so-called special sciences are to be admitted, they should be, in principle, connected to the

observational concepts of physics. Driven by epistemological motives, the logical positivists aimed, in effect, at a
double reduction: the reduction of the language of the special sciences to the language of physics and the reduction
of the language of physics to the intersubjective observational thing-language, that is, the language that refers to
middle-sized material objects. But it soon became clear
that the theoretical concepts are 'open-ended': they have
excess content over and above their empirical manifes-
tations. This development discredited the second strand
of the double reductive project, but the first strand was
alive for many decades to come. For even if the language
of physics could not be reduced to the thing-language, it
was still considered possible to reduce the language of the
special sciences to the language of physics.

Further reading: Carnap (1932)

Universals: From Plato (429–347 BCE) and **Aristotle** on,
many philosophers thought that a number of philosophical problems (the general applicability of predicates,
the unity of the propositions, the existence of similarity among particulars, the generality of knowledge and
others) required positing a separate type of entity – the
universal – alongside the **particulars**. Philosophers who
are realists about universals take universals to be really
there in the world, as constituents of states-of-affairs. Universals are the features that several distinct particulars
share in common (e.g., redness or triangularity). They are
the **properties** and relations in virtue of which particulars
are what they are and resemble other particulars. They
are also the referents of predicates. For instance, whiteness is the universal in virtue of which all white things are
white (the property they share); it is also the referent of
the predicate 'is white'; and together with a particular, for
example, a piece of chalk, it constitutes the state-of-affairs

of this chalk being white. Universals are taken to be the repeatable and recurring features of nature. When we say, for instance, that two apples are both red, we should mean that the very same property (redness) is instantiated by the two particulars (the apples). The very idea that universals are entities in their own right leads to the problem of how they are related to particulars and how they bind with them in a state-of-affairs. Philosophers have posited the relation of instantiation: universals are instantiated in (or by) particulars. But this relation has not been properly explicated and has often been taken as primitive. In recent decades, universals have been employed to explain **laws of nature**. One prime reason for positing universals is the **truth-maker principle**. But this principle is not uncontroversial, especially when it comes to universals. **Quine,** for instance, resists the thesis that, since we can make true statements that involve predicates, we should be ontologically committed to the existence of properties as self-subsisting universals. He aims to account for the role that universals are supposed to play by other means, for example, in terms of sets or classes, which, unlike the universals, are supposed to have clear identity conditions. For instance, claims such as 'wisdom is a virtue' should be understood as follows: for all x, if x is wise then x is virtuous. This last claim does not imply the existence of universals. Instead, it claims that the class of wise things is a subclass of the virtuous things. Though there are many varieties of **nominalism,** they all unite in denying that universals are self-subsistence entities.

See **Laws of thinghood; Tropes**
Further reading: Armstrong (1989)

Unobservable entities: Entities, for example, electrons, or DNA molecules, that cannot be seen with the naked eye. They are posited as constituents of observable objects and/or as causes of their observable behaviour. Many scientific theories seem to assume their existence. Scientific realists take it that there are such entities (i.e., that the world has a deep unobservable structure) and that scientific theories describe their behaviour. Empiricists (but not all of them) have typically felt that positing unobservable entities is illegitimate, since their existence transcends what can be known directly from observation and experiment.

See **Constructive empiricism; Empiricism; Entity realism; Scientific realism**

Further reading: Psillos (1999)

Vaihinger, Hans (1852–1933): German philosopher, author of *The Philosophy of As If* (1911). He is the founder of **fictionalism**. He noted that what is meant by saying that matter consists of atoms is that matter must be treated *as if* it consisted of atoms. Though it is false that matter has atomic structure, Vaihinger argued that the as-if operator implies a decision to maintain formally the assumption that matter has atomic structure as a useful fiction. Hence, we may willingly accept falsehoods or fictions if this is useful for practical purposes or if we thereby avoid conceptual perplexities. We then act *as if* they were true or real.

See **Fictionalism, mathematical**

Further reading: Vaihinger (1911)

Validation vs vindication: Distinction pertaining to rules of **inference** or propositions introduced by **Feigl**. A rule or a proposition is validated if it is derived (or shown to be an instance of) a more fundamental rule or proposition. *Modus tollens*, for instance, can be validated by being shown to be an instance of *modus ponens*. A rule or a proposition is vindicated if it is shown that it can successfully lead to the fulfilment of an aim (normally an aim for which the rule or the proposition is chosen or designed). For instance, deductive inferential rules can be vindicated by showing that they can successfully meet the aim of truth-preservation: they do not lead from true premises to false conclusions. Clearly, not all rules of inference can be validated; some must be taken as basic: they validate the others. But, on Feigl's view, even basic rules of **inference** can be vindicated. Following **Reichenbach**, Feigl argued that basic inductive rules, like the **straight rule of induction**, can be vindicated – they can be shown to be successful in meeting the aim of correct prediction of the future. Vindication amounts to a kind of pragmatic, or instrumental, **justification**: a rule is justified by being the best means to a certain end.

Further reading: Feigl (1950); Salmon (1967)

van Fraassen, Bas C. (born 1941): American philosopher of science, author of *The Scientific Image* (1980) and *The Empirical Stance* (2002). He has defended **constructive empiricism** as an alternative to **scientific realism**. He has also tried to develop an image of science that does away with **laws of nature** and the concept of **confirmation** of hypotheses, while taking a pragmatic approach to **explanation**. More recently, he has tried to develop a consistent version of **empiricism** in the sense that, given that empiricism denies the legitimacy of metaphysics, it should avoid being itself a metaphysical thesis – expressing a **belief** about the limits of experience. Van Fraassen has taken empiricism to be an epistemic policy (stance) that respects science but also criticises it in so far as it aims to offer explanations of the phenomena by positing **unobservable entities**. He has developed a new epistemology that is supposed to be in the service of empiricism, while avoiding **foundationalism** and **naturalism**. Van Fraassen has made substantial contributions to many areas of the philosophy of science, including the **semantic view of theories** and the interpretation of **quantum mechanics**.

See **Acceptance; Empirical adequacy; Explanation, pragmatics of; Models; Voluntarism**

Further reading: Ladyman (2002); van Fraassen (1980)

Verifiability: A statement is verifiable if its **truth** can be established in experience. Advocates of **logical positivism** took verifiability as a criterion of cognitive significance: those statements are meaningful whose truth can be verified. In slogan form: meaning is the method of verification. The logical positivists mobilised this criterion to show that statements of metaphysics are meaningless. There are several options as to how exactly to understand verifiability and these lead to different results as to what statements are meaningful. In the thought of logical positivists, the concept moved from a strict sense of provability on the basis of experience to the much more liberal sense of confirmability. In any case, as a criterion of meaning, verifiability has failed to deliver the goods. On its basis, apart from metaphysical statements, many ordinary scientific assertions, for example, those that express universal **laws of nature**, end up being meaningless. Besides, even nonsensical statements can be rendered meaningful by this criterion. Some philosophers objected to the verifiability criterion of meaning along the following lines: since it is not an analytic truth, if it is meaningful it should itself be verifiable. But it is not!

See **Protocol sentences; Verificationism**
Further reading: Ayer (1959)

Verificationism: Cluster of philosophical views united by the thought that the possibility of verification in experience is the sole criterion for attributing meaning, **justification**, **truth** and the like. The logical positivists favoured a verificationist criterion of meaningfulness, by deeming meaningless whatever proposition was not verifiable. The verificationism associated with **pragmatism** has mostly to do with the claim that the justification of a **belief** is a matter of the difference it makes in experience, and ultimately of its usefulness in inquiry. Modern verificationism, associated with **Dummett** and his followers concerns mostly the concept of truth, claiming that the concept of truth should be such that it cannot be meaningfully applied to propositions that cannot be verified.

See **Verifiability**
Further reading: Misak (1995)

Verisimilitude: Concept introduced by **Popper** to capture the idea that false theories may nonetheless be close to the **truth**. In particular, existing **scientific theories** may be false but they may also be more verisimilar (i.e., closer to the truth) than their predecessors. Popper offered a formal definition of comparative verisimilitude. Its gist is that a theory A is less verisimilar than a theory B if and only if (1) the contents of the theories are comparable and (2) *either* A has less truths than B and B has no more falsehoods than A; *or* A has no more truths than B and B has less falsehoods than A. It turned out that this account is deeply flawed. If we try to get a more verisimilar theory B from a false theory A by *adding* more truths to A, we also add more falsehoods to B, which are not falsehoods of A. Similarly, if we try to get a more verisimilar theory B from a false theory A by *subtracting* falsehoods from A, we also subtract truths from A, which are not truths of B.

See **Truthlikeness**

Further reading: Niiniluoto (1987)

Vienna Circle: Philosophical circle advocating **logical positivism** formed around **Schlick** in Vienna between 1922 and 1938. It started its meetings after the arrival of Schlick at the University of Vienna, essentially stopped its functioning after Schlick's assassination (in 1936) and was disbanded after the annexation of Austria by Nazi Germany (in 1938). Members of the Circle included: **Carnap, Neurath, Feigl,** Friedrich Waismann (1896–1959), Philip Frank (1884–1966), Kurt Goedel (1906–1978) and Hans Hahn (1879–1934). The Circle had close links with the Society for Empirical Philosophy in Berlin (members of which were **Reichenbach,** Kurt Grelling (1886–1942) and **Hempel).** The Circle made its first public appearance in 1929 with a manifesto, entitled *Vienna Circle: Its Scientific Outlook*. This associated the Circle with the doctrines of empiricist philosophers such as **Hume** and **Mach,** conventionalist philosophers of science such as **Poincaré** and **Duhem,** and logicians from **Leibniz** to **Russell.** The critique of metaphysics was emblematic of the Circle. The Circle edited *Erkenntnis*, a renowned philosophical journal (between 1930 and 1938) and organised a number of international congresses for scientific philosophy. After the dissolution of the Circle, *Erkenntnis* and the books of the members of the Circle were banned. Neurath and Waismann took refuge in England and Goedel in the USA. Though Wittgenstein was never a member of the Circle, his *Tractatus Logico-Philosophicus* exerted immense influence on their thinking. **Carnap** joined the Circle in

1926 and soon became one of its leading figures until he left for the USA in 1935. **Popper** was never a member of the Circle, but had regular discussions with its members.

See **Protocol sentences; Unity of science; Verifiability**
Further reading: Ayer (1959)

Vitalism: The doctrine that life is explained by the presence of vital forces; hence no mechanistic explanation of life is possible. It became popular in the beginning of the twentieth century as an anti-reductionist view in biology that relied one some notion of **emergence** to explain life. The French philosopher Henri Bergson (1859–1941) posited the presence of a vital force (*élan vital*), distinct from inert matter, to act as a principle for the organisation of bits of matter into a living organism. Vitalism fell into disrepute because it was taken to be in conflict with the principle of conservation of energy.

See **Mechanism; Reduction**
Further reading: Sober (1993)

Voluntarism: The view that having a **belief** is something that a person does voluntarily and can control. It is also equated with the kindred view that there can be reasons to believe that are *not* evidential. So one can come to believe that p (i.e., one can decide to believe that p) on the basis of reasons that are not related to the **probability** of p being true. There is a rather decisive argument against voluntarism: it is (pragmatically) incoherent to say that I believe at will. Belief aims (constitutively) at **truth**. If I could acquire a belief at will, I could acquire it whether it was true or not. Being *my* belief, I take it to be *true*. But I also *know* that my belief could be acquired whether it was true or not. Hence, I am (pragmatically) incoherent. I am saying: I believe that p (is true) but I believe that p whether it is true or not. Note that the incoherence noted above is not a formal contradiction, as can easily be seen if we replace the 'I' with a 'she': 'she believes that p (is true) but she believes that p whether it is true or not' might well be true. Yet, when this sentence is uttered by *me*, it is (pragmatically) incoherent. A form of voluntarism is a central plank of **van Fraassen**'s new epistemology.

See **Pascal's wager**

Further reading: van Fraassen (2002); Williams (1973)

Von Wright, Georg Henrik (1916–2003): Finnish philosopher, student of and successor to Wittgenstein in Cambridge as Professor of Philosophy in 1948. He worked on many central areas of the philosophy of science, most notably **causation, induction** and **probability**. He authored *A Treatise on Induction and Probability* (1951) and *Explanation and Understanding* (1971). He advanced an account of **causation** based on human action and on the possibility of manipulation.

Further reading: von Wright (1971)

W

Watkins, John (1924–1999): British philosopher of science, follower of **Popper** and successor of **Lakatos** at the London School of Economics. He authored *Science and Scepticism* (1984). He defended a largely Popperian view of science (though he wanted to do away with the notion of **verisimilitude**), and tried to meet the challenges faced by the notion of **corroboration**. He took it that theories should maximise testable content, explanatory depth and theoretical unity.

Further reading: Watkins (1984)

Whewell, William (1794–1866): English historian and philosopher of science, a central figure of Victorian science. He was among the founders of the British Association for the Advancement of Science, a fellow of the Royal Society and Master of Trinity College, Cambridge. Whewell coined the word 'scientist' in 1833. He took from **Kant** the view that ideas (or concepts) are necessary for experience in that only through them facts can be bound together. He noted, for instance, that **induction** requires a 'new mental element'. The concept of elliptical orbit, for instance, is not already there in the astronomical data that Kepler employed, but was a new mental element added by Kepler. But, unlike Kant, Whewell thought that history (and the history of science in particular) had a key role to play in understanding science and its philosophy. This role was analysed in *The Philosophy of the Inductive Sciences, Founded Upon Their History* (1840). According to Whewell, each science grows through three stages. It begins with a *prelude* in which a mass of unconnected facts is collected. It then enters an *inductive epoch* in which useful theories put order to these facts through the creative role of the scientists – an act of *colligation*. Finally, a *sequel* follows where the successful theory is extended, refined and applied. Whewell strongly emphasised the role of hypotheses in science. These hypotheses can be proved true, he thought, by the *consilience of inductions*. This is another expression coined by him. He meant it to refer to the theoretical **unification** which occurs when a theory explains data of a kind different from those it was initially introduced to explain, that is, when a theory unifies hitherto unrelated empirical domains. He thought that the consilience of inductions is a *criterion* of **truth**, a 'stamp of truth' or as he put it 'the point where truth resides'. His contemporary **Mill** claimed that no predictions could *prove* the truth of a theory and was involved in a heated debate with Whewell on this matter. Among Whewell's other major works is *History of the Inductive Sciences, from the earliest to the present time* (1847).

Further reading: Whewell (1989)

William of Ockham see **Ockham, William of**

Zahar, Elie (born 1937): Lebanese-born British philosopher of science, student of **Lakatos** and one of the most eloquent defenders of the Methodology of Scientific Research Programmes, which he applied to the case of the transition from Newtonian mechanics to **Einstein**'s theory of relativity. He is the author of *Einstein's Revolutions: A Study in Heuristics* (1989). He has worked on **Poincaré**'s philosophy of science and has defended **structural realism**. He is also known for his work on **ad hocness** and **novel predictions**.

Further reading: Zahar (1989; 2001)

Bibliography

Achinstein, Peter (2001), *The Book of Evidence*, New York: Oxford University Press.

Achinstein, Peter (ed.) (2005), *Scientific Evidence: Philosophical Theories and Applications*, Baltimore: Johns Hopkins University Press.

Albert, David (1992), *Quantum Mechanics and Experience*, Cambridge, MA: Harvard University Press.

Albert, David (2000), *Time and Chance*, Cambridge, MA: Harvard University Press.

Alcoff, Linda, and Elizabeth Potter (eds) (1993), *Feminist Epistemologies*, New York: Routledge.

Arabatzis, Theodore (2006), *Representing Electrons*, Chicago: University of Chicago Press.

Aristotle (1993), *Posterior Analytics*, Oxford: Clarendon Press.

Armstrong, D. M. (1983), *What Is a Law of Nature?*, Cambridge: Cambridge University Press.

Armstrong, D. M. (1989), *Universals: An Opinionated Introduction*, Boulder, CO: Westview Press.

Armstrong, D. M. (2004), *Truth and Truthmakers*, Cambridge: Cambridge University Press.

Ayer, A. J. (1936), *Language, Truth, and Logic*, Oxford: Oxford University Press.

Ayer, A. J. (ed.) (1959), *Logical Positivism*, New York: Free Press.

Baghramian, Maria (2004), *Relativism*, London: Routledge.

Bacon, Francis (1620), *The New Organon*, ed. Fulton H. Anderson, London: MacMillan, 1960.

Batterman, Robert (2001), *The Devil in the Details: Asymptotic Reasoning in Explanation, Reduction, and Emergence*, Oxford: Oxford University Press.

Bealer, George (1987), 'The Philosophical Limits of Scientific Essentialism', *Philosophical Perspectives* 1, 289–365.

Berkeley, George (1977), *The Principles of Human Knowledge with Other Writings*, London: Fontana Collins.

Bird, Alexander (1998), *Philosophy of Science*, Montreal: McGill-Queen's University Press.

Bird, Alexander (2000), *Thomas Kuhn*, Princeton: Princeton University Press.

Blackburn, Simon (1993), *Essays in Quasi-Realism*, New York: Oxford University Press.

Bloor, David (1991), *Knowledge and Social Imagery*, 2nd edn, Chicago: University of Chicago Press.

Boghossian, Paul (1996), 'Analyticity Reconsidered', *Noûs* 30, 360–92.

BonJour, Laurence (1985), *The Structure of Empirical Knowledge*, Cambridge, MA: Harvard University Press.

Boyd, Richard (1981), 'Scientific Realism and Naturalistic Epistemology', in P. D. Asquith and T. Nickles (eds), *PSA 1980*, Vol. 2, East Lansing: Philosophy of Science Association.

Boyd, R., P. Gasper and J. D. Trout (eds) (1991), *The Philosophy of Science*, Cambridge, MA: MIT Press.

Boyle, Robert (1979), *Selected Philosophical Papers of Robert Boyle*, ed. M. A. Stewart, Manchester: Manchester University Press.

Bridgman, P. W. (1927), *The Logic of Modern Physics*, New York: MacMillan.

Brown, James Robert (1991), *Laboratory of the Mind: Thought Experiments in the Natural Sciences*, London: Routledge.

Campbell, D. T. (1974), 'Evolutionary Epistemology', in P. A. Schilpp (ed.), *The Philosophy of Karl Popper*, La Salle, IL: Open Court, pp. 413–63.

Campbell, Keith (1990), *Abstract Particulars*, Oxford: Blackwell.

Carnap, Rudolf (1928), *The Logical Structure of the World*, trans. R. George Berkeley: University of California Press, 1967.

Carnap, Rudolf (1932), *The Unity of Science*, trans. M. Black, London: Kegan Paul.

Carnap, Rudolf (1934), *The Logical Syntax of Language*, trans. A. Smeaton, London: Kegan Paul, 1937.

Carnap, Rudolf (1936), 'Testability and Meaning', *Philosophy of Science* 3, 419–71.

Carnap, Rudolf (1950a), 'Empiricism, Semantics and Ontology', *Revue Intérnationale de Philosophie* 4, 20–40.

Carnap, Rudolf (1950b), *Logical Foundations of Probability*, Chicago: The University of Chicago Press.

Carnap, Rudolf (1956), 'The Methodological Character of Theoretical Concepts', *Minnesota Studies in the Philosophy of Science* 1, pp. 38–76.

Carnap, Rudolf (1974), *An Introduction to the Philosophy of Science*, New York: Basic Books.

Carroll, John W. (1994), *Laws of Nature*, Cambridge: Cambridge University Press.

Cartwright, Nancy (1983), *How the Laws of Physics Lie*, Oxford: Clarendon Press.

Cartwright, Nancy (1999), *The Dappled World*, Cambridge: Cambridge University Press.

Cassirer, Ernst (1910), *Substance and Function*, trans. William Curtis Swabey and Marie Curtis Swabey, Chicago: Open Court, 1923.

Chisholm, Roderick M. (1982), *The Foundations of Knowing*, Minneapolis: University of Minnesota Press.

Clark, Peter, and Catherine Hawley (eds) (2003), *Philosophy of Science Today*, Oxford: Clarendon Press.

Cohen, I. Bernard (1985), *The Birth of a New Physics*, London: Penguin.

Colyvan, Mark (2001), *The Indispensability of Mathematics*, New York: Oxford University Press.

Comte, Auguste (1913), *The Positive Philosophy of Auguste Comte*, trans. Harriet Martineau, London: Bell.

Cottingham, John (1984), *Rationalism*, London: Paladin Books.

Da Costa, Newton C. A. and Steven French (2003), *Science and Partial Truth*, New York: Oxford University Press.

Davidson, Donald (1980), *Essays on Actions and Events*, Oxford: Clarendon Press.

de Regt, H. W. (2005), 'Scientific Realism in Action: Molecular Models and Boltzmann's *Bildtheorie*', *Erkenntnis* 63, 205–30.

Descartes, René (1644), *Principles of Philosophy*, in *The Philosophical Writings of Descartes*, Vol. 1, trans. John Cottingham, Robert Stootfoff and Dugald Murdoch, Cambridge: Cambridge University Press, 1985.

Devitt, Michael (1997), *Realism and Truth*, with a new 'Afterword', 1st edn 1984, Princeton: Princeton University Press.

Devitt, Michael, and Kim Sterelny (1987), *Language and Reality*, Oxford: Blackwell.

deVries, Willem (2005), *Wilfrid Sellars*, Chesham: Acumen.

Dowe, Phil (2000), *Physical Causation*, Cambridge: Cambridge University Press.

Ducasse, C. J. (1969), *Causation and Types of Necessity*, New York: Dover.

Duhem, Pierre (1906), *The Aim and Structure of Physical Theory*, trans. P. Wiener, Princeton: Princeton University Press, 1954.

Dummett, Michael (1991), *The Logical Basis of Metaphysics*, Cambridge, MA: Harvard University Press.

Earman, John (1986), *A Primer on Determinism*, Dordrecht: Reidel.

Earman, John (1989), *World Enough and Space-Time: Absolute Versus Relational Theories of Space and Time*, Cambridge, MA: MIT Press.

Earman, John (1992), *Bayes or Bust? A Critical Examination of Bayesian Confirmation Theory*, Cambridge, MA: MIT Press.

Eells, Ellery (1991), *Probabilistic Causality*, Cambridge: Cambridge University Press.

Ellis, Brian (2001), *Scientific Essentialism*, Cambridge: Cambridge University Press.

Engel, Morris S. (2000), *With Good Reason: An Introduction to Informal Fallacies*, Bedford: St. Martin's.

Feigl, Herbert (1950), '*De principiis non disputandum . . .* ?', in Max Black (ed.), *Philosophical Analysis, a Collection of Essays*, Ithaca: Cornell University Press, pp. 119–56.

Feigl, Herbert (1981), *Inquiries and Provocations: Selected Writings 1929–1974*, Dordrecht: Reidel.

Feyerabend, Paul K. (1975), *Against Method*, London: New Left Books.

Field, Hartry (1980), *Science without Numbers*, Oxford: Blackwell.

Fine, Arthur (1986), *The Shaky Game*, Chicago: University of Chicago Press.

Fine, Kit (2002), *The Limits of Abstraction*, Oxford: Clarendon Press.

Fitch G. W. (2004), *Saul Kripke*, Chesham: Acumen.

Fodor, Jerry (1974), 'Special Sciences, or the Disunity of Science as a Working Hypothesis', *Synthese* 28, 97–115.

Fodor, Jerry (1998), *Concepts*, New York: Clarendon Press.

Fodor, Jerry, and Ernest Lepore (1992), *Holism: A Shopper's Guide*, Oxford: Blackwell.

Forster, Malcolm, and Elliott Sober (1994), 'How to Tell When Simpler, More Unified, or Less *Ad Hoc* Theories Will Provide More Accurate Predictions', *British Journal for the Philosophy of Science* 45, 1–36.

Frege, Gottlob (1884), *The Foundations of Arithmetic*, trans. J. L. Austin, Evanston, IL: Northwestern University Press, 1980.

Friedman, Michael (1999), *Reconsidering Logical Positivism*, Cambridge: Cambridge University Press.

Galileo Galilei (1638), *Two Dialogues Concerning the Two New Sciences*, Encyclopaedia Britannica, 1952.

Galison, Peter (1987), *How Experiments End*, Chicago: University of Chicago Press.

Garfinkel, Alan (1981), *Forms of Explanation*, New Haven: Yale University Press.

Gascoigne, Neil (2002), *Scepticism*, Chesham: Acumen.

Giedymin, Jerzy (1982), *Science and Convention*, Oxford: Pergamon Press.

Giere, Ronald (1988), *Explaining Science: A Cognitive Approach*, Chicago: University of Chicago Press.

Giere, Ronald (1999), *Science without Laws*, Chicago: University of Chicago Press.

Giere, Ronald, and Alan Richardson (eds) (1996), *Origins of Logical Empiricism*, Minneapolis: University of Minnesota Press.

Gillies, Donald (2000), *Philosophical Theories of Probability*, London: Routledge.

Glennan, Stuart (2002), 'Rethinking Mechanical Explanation', *Philosophy of Science* 69, S342–S353.

Glymour, Clark (1980), *Theory and Evidence*, Princeton: Princeton University Press.

Godfrey-Smith, Peter (2003), *Theory and Reality: An Introduction to the Philosophy of Science*, Chicago: University of Chicago Press.

Goldman, A. I. (1986), *Epistemology and Cognition*, Cambridge, MA: Harvard University Press.

Goodman, Nelson (1954), *Fact, Fiction and Forecast*, Cambridge, MA: Harvard University Press.

Gower, Barry (1998), *Scientific Method: An Historical and Philosophical Introduction*, London: Routledge.

Grünbaum, Adolf (1973), *Philosophical Problems of Space and Time*, 2nd enlarged edn, Dordrecht: Reidel.

Guyer, Paul (ed.) (1992), *The Cambridge Companion to Kant*, Cambridge: Cambridge University Press.

Hacking, Ian (1965), *The Logic of Statistical Inference*, Cambridge: Cambridge University Press.

Hacking, Ian (1983), *Representing and Intervening*, Cambridge: Cambridge University Press.

Hájek, Alan (2003), 'Waging War on Pascal's Wager', *Philosophical Review* 112, 27–56.

Hale, Bob (1987), *Abstract Objects*, Oxford: Blackwell.

Hanson, Norwood Russell (1958), *Patterns of Discovery*, Cambridge: Cambridge University Press.

Harding, Sandra (1986), *The Science Question in Feminism*, Ithaca, NY: Cornell University Press.

Harman, Gilbert (1986), *Change in View: Principles of Reasoning*, Cambridge, MA: MIT Press.

Harré, Rom, and E. H. Madden (1975), *Causal Powers: A Theory of Natural Necessity*, Oxford: Blackwell.

Heil, John (2003), *From an Ontological Point of View*, Oxford: Clarendon Press.

Hempel, Carl (1965), *Aspects of Scientific Explanation and Other Essays in the Philosophy of Science*, New York: Free Press.

Hertz, Heinrich (1894), *The Principles of Mechanics Presented in a New Form*, New York: Dover Publications, 1955.

Hesse, M. B. (1966), *Models and Analogies in Science*, Notre Dame: University of Notre Dame Press.

Hilbert, David (1899), *The Foundations of Geometry*, trans. Leo Unger, Chicago: Open Court Publishing Company, 1971.

Hitchcock, Christopher (ed.) (2004), *Contemporary Debates in the Philosophy of Science*, Oxford: Blackwell.

Holyoak, Keith J. and Paul Thagard (1995), *Mental Leaps*, Cambridge, MA: MIT Press.

Horwich, Paul (1987), *Asymmetries in Time*, Cambridge, MA: MIT Press.

Horwich, Paul (1998a), *Meaning,* Oxford: Oxford University Press.

Horwich, Paul (1998b), *Truth,* 2nd edn, Oxford: Oxford University Press.

Howson, Colin (2000), *Hume's Problem*, New York: Oxford University Press.

Howson, Colin, and Peter Urbach (2006), *Scientific Reasoning: The Bayesian Approach*, 3rd edn, Chicago: Open Court Publishing Company.

Huemer, Michael (2001), *Skepticism and the Veil of Perception*, Lanham, MD: Rowman & Littlefield.

Hull, David (1988), *Science as a Process: An Evolutionary Account of the Social and Conceptual Development of Science*, Chicago: University of Chicago Press.

Hull, David, and Michael Ruse (eds) (1998), *The Philosophy of Biology*, Oxford: Oxford University Press.

Hume, David (1739), *A Treatise of Human Nature*, ed. L. A. Selby-Bigge 1888, 2nd edn, ed. P. H. Nidditch, Oxford: Clarendon Press, 1978.

Humphreys, Paul (1989), *The Chances of Explanation*, Princeton: Princeton University Press.

James, William (1897), *The Will to Believe and Other Essays in Popular Philosophy*, Cambridge, MA: Harvard University Press, 1979.

Kant, Immanuel (1787), *Critique of Pure Reason*, trans. Norman Kemp Smith, New York: St. Martin's Press, 1965.

Keynes, John Maynard (1921), *A Treatise on Probability*, London: Macmillan.

Kim, Jaegwon (1993), *Supervenience and Mind*, Cambridge: Cambridge University Press.

Kincaid, Harold (1996), *Philosophical Foundations of the Social Sciences: Analyzing Controversies in Social Research*, Cambridge: Cambridge University Press.

Kirkham, R. L. (1992), *Theories of Truth: A Critical Introduction*, Cambridge, MA: MIT Press.

Kitcher, Philip (1989), 'Explanatory Unification and Causal Structure', *Minnesota Studies in the Philosophy of Science* 13, 410–505.

Kitcher, Philip (1993), *The Advancement of Science*, Oxford: Oxford University Press.

Klein, Peter (1984), *Certainty: A Refutation of Scepticism*, Minneapolis: University of Minnesota Press.

Kneale, William (1949), *Probability and Induction*, Oxford: Clarendon Press.

Koertge, Noretta (ed.) (1998), *A House Built on Sand: Exposing Postmodernist Myths about Science*, Oxford: Oxford University Press.

Kripke, Saul (1980), *Naming and Necessity*, Oxford: Blackwell.

Kuhn T. S. (1957), *The Copernican Revolution*, Cambridge, MA: Harvard University Press.

Kuhn, T. S. (1962), *The Structure of Scientific Revolutions*, 2nd enlarged edn, 1970, Chicago: University of Chicago Press.

Kyburg, Henry E. (1974), *The Logical Foundations of Statistical Inference*, Dordrecht: Reidel.

Ladyman, James (2002), *Understanding Philosophy of Science*, London: Routledge.

Lakatos, Imre (1970), 'Falsification and the Methodology of Scientific Research Programmes', in Imre Lakatos and Alan Musgrave (eds), *Criticism and the Growth of Knowledge*, Cambridge: Cambridge University Press, pp. 91–196.

Lange, Marc (2000), *Natural Laws in Scientific Practice*, Oxford: Oxford University Press.

Lange, Marc (2002), *An Introduction to the Philosophy of Physics*, Oxford: Blackwell.

Langton, Rae, and David Lewis (1998), 'Defining "Intrinsic" ', *Philosophy and Phenomenological Research* 58, 333–45.

Laplace, Pierre Simon (1814), *A Philosophical Essay on Probabilities*, New York: Dover, 1951.

Laudan, Larry (1996), *Beyond Positivism and Relativism*, Boulder: Westview Press.

Leibniz, Gottfried (1973), *Discourse on Metaphysics, Correspondence with Arnauld, Monadology*, trans. G. Montgomery, Chicago: Open Court Publishing Company.

Le Poidevin, Robin, and Murray MacBeath (eds) (1993), *The Philosophy of Time*, Oxford: Oxford University Press.

Leplin, Jarrett (1997), *A Novel Defence of Scientific Realism*, Oxford: Oxford University Press.

Lewis, David (1973a), 'Causation', *Journal of Philosophy* 70, 556–67.

Lewis, David (1973b), *Counterfactuals*, Cambridge, MA: Harvard University Press.

Lewis, David (1980), 'A Subjectivist's Guide to Objective Chance', in R. C. Jeffrey (ed.), *Studies in Inductive Logic and Probability Vol. II*, Berkeley: Berkeley University Press, pp. 63–93.

Lewis, David (1986), 'Causal Explanation', *Philosophical Papers, Vol. II*, Oxford: Oxford University Press, pp. 214–40.

Lewis, David (1999), *Papers in Metaphysics and Epistemology*, Cambridge: Cambridge University Press.

Lipton, Peter (2004), *Inference to the Best Explanation*, 2nd edn, London: Routledge.

Locke, John (1689), *An Essay Concerning Human Understanding*, Oxford: Clarendon Press, 1975.

Loewer, Barry (1996), 'Humean Supervenience', *Philosophical Topics* 24, 101–26.

Longino, Helen (1990), *Science as Social Knowledge*, Princeton: Princeton University Press.

Losee, John (2001), *A Historical Introduction to the Philosophy of Science*, Oxford: Oxford University Press.

Mach, Ernst (1910), *Popular Scientific Lectures*, Chicago: Open Court.

Machamer, Peter, Lindley Darden and Carl Craver (2000), 'Thinking about Mechanisms', *Philosophy of Science* 67, 1–25.

Mackie, J. L. (1974), *The Cement of the Universe: A Study of Causation*, Oxford: Clarendon Press.

McLaughlin, Brian (1992), 'The Rise and Fall of British Emergentism', in Ansgar Beckerman, Has Flohr and Jaewgon Kim (eds), *Emergence or Reduction?*, Berlin: De Gruyter, pp. 49–93.

McMullin, Ernan (1985), 'Galilean Idealisation', *Studies in History and Philosophy of Science* 16, 247–73.

McMullin, Ernan (1992), *The Inference That Makes Science*, Milwaukee: Marquette University Press.

Maher, Patrick (1993), *Betting on Theories*, Cambridge: Cambridge University Press.

Malebranche, Nicolas (1674–5), *The Search After Truth*, trans. Thomas M. Lennon and Paul J. Olscamp, Cambridge: Cambridge University Press, 1997.

Maxwell, Grover (1962), 'The Ontological Status of Theoretical Entities', *Minnesota Studies in the Philosophy of Science* 3, 3–27.

Maxwell, James Clerk (1890), *The Scientific Papers of James Clerk Maxwell*, ed. W. D. Niven, vols 1 and 2, New York: Dover Publications.

Mayo, Deborah G. (1996), *Error and the Growth of Experimental Knowledge*, Chicago: University of Chicago Press.

Mellor, D. H. (1991), *Matters of Metaphysics*, Cambridge: Cambridge University Press.

Mellor, D. H. (1995), *The Facts of Causation*, London: Routledge.

Mill, John Stuart (1911), *A System of Logic: Ratiocinative and Inductive*, London: Longmans, Green.

Miller, D. W. (1994), *Critical Rationalism: A Restatement and Defence*, Chicago: Open Court.

Misak, Cheryl J. (1995), *Verificationism: Its History and Prospects*, London: Routledge.

Morgan, Mary, and Margaret Morrison (1999) (eds), *Models as Mediators: Perspectives on Natural and Social Science*, Cambridge: Cambridge University Press.

Morrison, Margaret (2000), *Unifying Scientific Theories*, Oxford: Oxford University Press.

Mumford, Stephen (1998), *Dispositions*, Oxford: Clarendon Press.

Mumford, Stephen (2004), *Laws in Nature*, London: Routledge.

Murdoch, Dugald (1987), *Niels Bohr's Philosophy of Physics*, Cambridge: Cambridge University Press.

Musgrave, Alan (1999), *Essays on Realism and Rationalism*, Amsterdam: Rodopi.

Nagel, Ernst (1960), *The Structure of Science*, 2nd edn, Indianapolis: Hackett, 1979.

Nagel Ernest (1977), 'Teleology Revisited', *Journal of Philosophy* 75, 261–301.

Neurath, Otto (1983), *Philosophical Papers 1913–1946*, Dordecht: Reidel.

Newton-Smith, W. H. (1981), *The Rationality of Science*, London: Routledge & Kegan Paul.

Nicod, Jean (1969), *Geometry and Induction*, London: Routledge & Kegan Paul.

Niiniluoto, Ilkka (1987), *Truthlikeness*, Dordrecht: Reidel.

Nola, Robert (2003), *Rescuing Reason: A Critique of Anti-rationalist Views of Science and Knowledge*, Dordrecht: Kluwer.

Nola, Robert, and Howard Sankey (2000), 'A Selective Survey of Theories of Scientific Method', in Robert Nola and Howard Sankey (eds), *After Popper, Kuhn and Feyerabend: Recent Issues in Theories of Scientific Method*, Dordrecht: Kluwer Academic Publishers, 2000, pp. 1–65.

Nolan, Daniel (2005), *David Lewis*, Chesham: Acumen.

Nozick, Robert (1993), *The Nature of Rationality*, Princeton: Princeton University Press.

Nozick, Robert (2001), *Invariances*, Harvard: Harvard University Press.

Ockham, William of (1990) *Philosophical Writings: A Selection*, Indianapolis: Hackett Publishing Company.

Oddie, Graham (1986), *Likeness to Truth*, Dordrecht: Reidel.

Okasha, Samir (2001), *Philosophy of Science: A Very Short Introduction*, Oxford: Oxford University Press.

Orenstein, Alex (2002), *W. V. Quine*, Chesham: Acumen.

Papineau, David (1993), *Philosophical Naturalism*, Oxford: Blackwell.

Papineau, David (ed.) (1997), *The Philosophy of Science*, Oxford: Oxford University Press.

Peirce, C. S. (1957), *Essays in the Philosophy of Science*, ed. V. Tomas, New York: The Liberal Arts Press.

Plantinga, Alvin (1993), *Warrant: The Current Debate*, Oxford: Oxford University Press.

Poland, John (1994), *Physicalism: The Philosophical Foundations*, Oxford: Clarendon Press.

Poincaré, Henri (1902), *Science and Hypothesis*, New York: Dover Publications, 1905.

Pollock, John (1986), *Contemporary Theories of Knowledge*, Savage, MD: Rowan & Littlefield.

Popper, Karl (1959), *The Logic of Scientific Discovery*, London: Hutchinson.

Popper, Karl (1963), *Conjectures and Refutations*, 3rd edn rev., London: Routledge & Kegan Paul, 1969.

Preston, John (1997), *Feyerabend: Philosophy, Science and Society*, Cambridge: Polity Press.

Price, Huw (1996), *Time's Arrow and Archimedes' Point*, Oxford: Oxford University Press.

Psillos, Stathis (1999), *Scientific Realism: How Science Tracks Truth*, London: Routledge.

Psillos, Stathis (2002), *Causation and Explanation*, Chesham: Acumen.

Putnam, Hilary (1978), *Meaning and the Moral Sciences*, London: Routledge & Kegan Paul.

Putnam, Hilary (1981), *Reason, Truth and History*, Cambridge: Cambridge University Press.

Putnam, Hilary, and Paul Oppenheim (1958), 'Unity of Science as a Working Hypothesis', *Minnesota Studies in the Philosophy of Science* 2, pp. 3–36.

Pyle Andrew (1995), *Atomism and Its Critics: Democritus to Newton*, Bristol: Thoemmes.

Quine, W. v. O. (1951), 'Two Dogmas of Empiricism', *The Philosophical Review* 60, 20–43.

Quine, W. v. O. (1953), 'On What There Is', *From a Logical Point of View*, Cambridge, MA: Harvard University Press.

Quine, W. v. O. (1960), *Word and Object*, Cambridge, MA: MIT Press.

Quine, W. v. O. (1966), *The Ways of Paradox and Other Essays*, Cambridge, MA: Harvard University Press.

Quine, W. v. O. (1969), 'Epistemology Naturalised', *Ontological Relativity and Other Essays*, Cambridge, MA: Harvard University Press.

Quine, W. v. O. (1975), 'On Empirically Equivalent Systems of the World', *Erkenntnis* 9, 313–28.

Quine, W. v. O. and J. S. Ullian (1978), *The Web of Belief*, New York: Random House.

Quinton, Anthony (1973), *The Nature of Things*, London: Routledge & Kegan Paul.

Railton, Peter (1978), 'A Deductive-Nomological Model of Probabilistic Explanation', *Philosophy of Science* 45, 206–26.

Ramsey Frank (1931), *The Foundations of Mathematics and Other Essays*, ed. R. B. Braithwaite, London: Routledge & Kegan Paul.

Redhead, Michael (1987), *Incompleteness, Nonlocality and Realism*, Oxford: Clarendon Press.

Reichenbach, Hans (1921), *The Theory of Relativity and A Priori Knowledge*, trans. Maria Reichenbach, Berkeley and Los Angeles: University of California Press, 1965.

Reichenbach, Hans (1938), *Experience and Prediction*, Chicago: University of Chicago Press.

Reichenbach, Hans (1949), *The Theory of Probability*, Berkeley: University of California Press.

Reichenbach, Hans (1951), *The Rise of Scientific Philosophy*, Berkeley: University of California Press.

Reichenbach, Hans (1958), *The Philosophy of Space and Time*, New York: Dover Publications.

Resnik, David (1998), *The Ethics of Science*, New York: Routledge.

Rorty, Richard (1982), *The Consequences of Pragmatism*, Minneapolis: University of Minnesota Press.

Rosenberg, Alexander (2000), *Philosophy of Science: A Contemporary Introduction*, London: Routledge.

Russell, Bertrand (1912), *The Problems of Philosophy*, Oxford: Oxford University Press.

Russell, Bertrand (1927), *The Analysis of Matter*, London: Routledge & Kegan Paul.

Sahlin, Nils-Eric (1990), *The Philosophy of F. P. Ramsey*, Cambridge: Cambridge University Press.

Sainsbury, Mark (1979), *Russell*, London: Routledge & Kegan Paul.

Sainsbury, Mark (1988), *Paradoxes*, Cambridge: Cambridge University Press.

Salmon, Wesley (1967), *The Foundations of Scientific Inference*, Pittsburgh: University of Pittsburgh Press.

Salmon, Wesley (1984), *Scientific Explanation and the Causal Structure of the World*, Princeton: Princeton University Press.

Salmon, Wesley (1989), *Four Decades of Scientific Explanation*, Minneapolis: University of Minnesota Press.

Salmon, Wesley, Richard C. Jeffrey and James G. Greeno (1971), *Statistical Explanation and Statistical Relevance*, Pittsburgh: University of Pittsburgh Press.

Sankey, Howard (1994), *The Incommensurability Thesis*, Aldershot: Avebury.

Schlick, Moritz (1918), *General Theory of Knowledge*, 2nd German edn, trans. A. E. Blumberg, Vienna and New York: Springer-Verlag, 1925.

Schlick, Moritz (1979), *Philosophical Papers*, 2 vols, Dordrecht: Reidel.

Sellars, Wilfrid (1963), *Science, Perception and Reality*, Atascadero, CA: Ridgeview 1991.

Shapiro, Stuart (1997), *Philosophy of Mathematics: Structure and Ontology*, Oxford: Oxford University Press.

Shoemaker, S. (1984), *Identity, Cause, and Mind*, Cambridge: Cambridge University Press.

Sklar, Lawrence (1974), *Space, Time and Spacetime*, Berkeley: University of California Press.

Sklar, Lawrence (1995), *Physics and Chance*, Cambridge: Cambridge University Press.

Skyrms, Brian (2000), *Choice and Chance*, 4th edn, Belmont, CA: Wadsworth.

Smart J. J. C. (1963), *Philosophy and Scientific Realism*, London: Routledge & Kegan Paul.

Sober, Elliott (1990), 'Let's Razor Ockham's Razor', in D. Knowles (ed.), *Explanation and Its Limits*, Royal Institute of Philosophy Supplementary Vol. 27, Cambridge: Cambridge University Press, pp. 73–94.

Sober, Elliott (1993), *The Philosophy of Biology*, Boulder: Westview Press.

Sober, Elliott (2002), 'Bayesianism – Its Scope and Limits', in Richard Swinburne (ed.), *Bayesianism*, Proceedings of the British Academy, vol. 113, Oxford: Oxford University Press, pp. 21–38.

Solomon, Miriam (2001), *Social Empiricism*, Cambridge, MA: MIT Press.

Sosa, Ernst, and Michael Tooley (eds) (1993), *Causation*, Oxford: Oxford University Press.

Stalker, Douglas (1994), *Grue! The New Riddle of Induction*, La Salle: Open Court.

Stegmuller, Wolfgang (1979), *The Structuralist View of Theories*, Berlin: Springer.

Sterelny, Kim, and Paul E. Griffiths (1999), *Sex and Death: An Introduction to the Philosophy of Biology*, Chicago: University of Chicago Press.

Stove, David (1991), *The Plato Cult and Other Philosophical Follies*, Oxford: Blackwell.

Stroud, Barry (1977), *Hume*, London: Routledge.

Suppe, Fred (1989), *Scientific Realism and Semantic Conception of Theories*, Urbana: University of Illinois Press.

Suppe, Fred (ed.) (1977), *The Structure of Scientific Theories*, 2nd edn, Urbana: University of Illinois Press.

Suppes, Patrick (1984), *Probabilistic Metaphysics*, Oxford: Blackwell.

Swinburne, Richard (1997), *Simplicity as Evidence of Truth*, Milwaukee: Marquette University Press.

Swinburne, Richard (ed.) (1974), *The Justification of Induction*, Oxford: Oxford University Press.

Tarski, Alfred (1944), 'The Semantic Conception of Truth', in L. Linsky (ed.), *Semantics and the Philosophy of Language*, Urbana: University of Illinois Press, 1970, pp. 13–47 – first appeared in *Philosophy and Phenomenological Research* 4, 341–76.

Tarski, Alfred (1969), 'Truth and Proof', *Scientific American* 220, 63–77.

Torretti, Roberto (1978), *Philosophy of Geometry from Riemann to Poincaré*, Dordrecht: Reidel.

Torretti, Roberto (1999), *The Philosophy of Physics*, New York: Cambridge University Press.

Uebel, Thomas (1992), *Overcoming Logical Positivism from Within*, Amsterdam: Rodopi.

Unger, Peter (1983), 'The Causal Theory of Reference', *Philosophical Studies* 43, 1–45.

Vaihinger, Hans (1911), *The Philosophy of 'As If'*, trans. C. K. Ogden, London: Routledge, 1924.

van Fraassen, Bas C. (1980), *The Scientific Image*, Oxford: Clarendon Press.

van Fraassen, Bas C. (1985), 'Empiricism in Philosophy of Science', in P. M. Churchland and C. A. Hooker (eds), *Images of Science*, Chicago: University of Chicago Press, pp. 245–308.

van Fraassen, Bas C. (2002), *The Empirical Stance*, New Haven and London: Yale University Press.

Vision, Gerald (2004), *Veritas: The Correspondence Theory and Its Critics*, Cambridge, MA: MIT Press.

von Mises Richard (1957), *Probability, Statistics and Truth*, rev. English edn, New York: Macmillan.

von Wright, G. H. (1971), *Explanation and Understanding*, London: Routledge & Kegan Paul.

Watkins, John (1984), *Science and Scepticism*, Princeton: Princeton University Press.

Weiner, Joan (2004), *Frege Explained*, Chicago: Open Court.

Weiss, Bernard (2002), *Michael Dummett*, Chesham: Acumen.

Whewell William (1989), *Theory of Scientific Method*, edited with an introduction by R. Butts, Indianapolis: Hackett.

Williams, Bernard (1973), *Problems of the Self*, Cambridge: Cambridge University Press.

Williams, Michael (2001), *Problems of Knowledge*, Oxford: Oxford University Press.

Wilkerson, T. E. (1995), *Natural Kinds*, Avebury: Ashgate Publishing Company.

Wilson, Margaret (1999), *Ideas and Mechanism*, Princeton: Princeton University Press.

Winkler, K. P. (1989), *Berkeley: An Interpretation*, Oxford: Clarendon Press.

Woodward, James (2003), *Making Things Happen: A Theory of Causal Explanation*, New York: Oxford University Press.

Worrall, John (1989), 'Structural Realism: The Best of Both Worlds', *Dialectica* 43, 99–124.

Wright, Crispin (1992), *Truth and Objectivity*, Cambridge, MA: Harvard University Press.

Wright, Larry (1976), *Teleological Explanations: An Etiological Analysis of Goals and Functions*, Berkeley: University of California Press.

Zahar, Elie (1989), *Einstein's Revolution*, La Salle: Open Court.

Zahar, Elie (2001), *Poincaré's Philosophy: From Conventionalism to Phenomenology*, La Salle: Open Court.